Lecture Notes in Computer Science   8420

Commenced Publication in 1973
Founding and Former Series Editors:
Gerhard Goos, Juris Hartmanis, and Jan van Leeuwen

For further volumes:
http://www.springer.com/series/8637

Abdelkader Hameurlain · Josef Küng
Roland Wagner (Eds.)

Transactions on Large-Scale Data- and Knowledge-Centered Systems XIII

 Springer

Editors-in-Chief
Abdelkader Hameurlain
IRIT
Paul Sabatier University
Toulouse
France

Josef Küng
Roland Wagner
FAW
University of Linz
Linz
Austria

ISSN 0302-9743   ISSN 1611-3349 (electronic)
ISBN 978-3-642-54425-5   ISBN 978-3-642-54426-2   (eBook)
DOI 10.1007/978-3-642-54426-2
Springer Heidelberg New York Dordrecht London

Library of Congress Control Number: 2014933393

Printed on acid-free paper

Springer is part of Springer Science+Business Media (www.springer.com)

Preface

In the last decade, we have witnessed a continuing data explosion generated by multiple data sources. These can be material (e.g., sensors), human (e.g., scientific activities), commercial, etc... The variety of data sources, the huge volumes of data and their high heterogeneity create new problems at two levels: methodological and engineering. In large-scale environments, data modelling, knowledge discovery, information filtering and efficient querying of data sources present an important and challenging issue.

This volume contains 6 fully revised selected regular papers. Its content covers a wide range of different hot topics in the field of data management, mainly: federated data sources, information filtering, web data clouding, query reformulation, package skyline queries, and SPARQL query processing over a LaV (Local-as-View) integration system.

We would like to express our thanks to the Editorial Board for thoroughly refereeing the submitted papers and ensuring the high quality of this volume. Special thanks go to Gabriela Wagner for her availability and her valuable work in the realization of this TLDKS volume.

January 2014

Abdelkader Hameurlain
Josef Küng
Roland Wagner

Editorial Board

Contents

Contents

Enabling a Package Query Paradigm on the Semantic Web: Model and Algorithms

Matthew Sessoms[✉] and Kemafor Anyanwu

Semantic Computing Research Lab, Department of Computer Science,
North Carolina State University, Raleigh, NC, USA
{mwsessom,kogan}@ncsu.edu
http://www.ncsu.edu

Abstract. The traditional search model of finding links on the Web is unsatisfactory for the increasingly complex tasks that seek to leverage the diverse, increasingly structured and semantically annotated data on the Web. A good example is when users seek to find collections or packages of resources that meet some constraints e.g., a collection of learning resources that cover some topics and have a good average rating or a collection of tourist attractions in a city such that total cost and total travel time for visiting all attractions meet the given constraints. For such queries, the goal is the return a set of constraint-qualifying collections or *packages*. However, using the traditional "set of links" query paradigm, such queries can only be satisfied by issuing multiple queries, reviewing answer lists and manually assembling packages to suit a user's desired constraints.

In this article, we introduce the concept of a *Package Query* for querying for resource combinations on the Semantic Web. In particular, we consider a frequent subclass of such queries *Skyline Package Queries*, in which multiple competing criteria are specified in the query so that the pareto-optimal set or skyline of packages are returned. In contrast to a few recent efforts on package queries on single relational models, fine-grained data models such as RDF include the challenge of computing the package skyline over multiple joins of ternary relations. We present four evaluation strategies involving different combinations of relational query operators and a new operator for Skyline Package Queries and different storage models for RDF data. A comparative evaluation of the algorithms over real world and synthetic-benchmark RDF datasets is presented.

Keywords: RDF · Package Skyline queries · Performance

1 Introduction

The Web has become a dominant knowledge source that informs a wide variety of technical and non-technical decisions. An increasing number of decision tasks being supported by Web data require more complex search paradigms than the

A. Hameurlain et al. (Eds.): TLDKS XIII, LNCS 8420, pp. 1–32, 2014.
DOI: 10.1007/978-3-642-54426-2_1, © Springer-Verlag Berlin Heidelberg 2014

mainstream "list of url matches" paradigm on the Web. Although there may be limitations to the degree of search complexity that is possible over unstructured content, the growing success of the Semantic Web offers the potential of leveraging its large amount of structured knowledge content for answering complex questions. An interesting querying use case that has broad applications is one where satisfying a user's query can be achieved only with combinations or "packages" of resources and not individual resources. Therefore, a query result is a list of resource combinations and not simply a list of resources. As a more concrete example, consider the following scenario.

Motivating Example (*e-learning*). A student would like to find e-learning resources from a collection of semantically annotated e-learning resources, e.g., [37], covering a set of topics on an upcoming test: relational model, algebra and calculus. Since creators of resources have the flexibility to modularize their content as they see fit, these topics may be covered by a single resource by one author, or by multiple resources by a different author, e.g., splitting content into two units: "data modeling" which covers the relational model, and then "querying" which covers relational algebra and calculus. Yet another content creator has a seperate content resource for each of the three topics. The implication is that in order to satisfy the user's query, some results, which match only portions of the answer, will need to be "packaged" into combinations that completely satisfy the query. Further, assuming that individual resources have ratings from reviews and maybe subscription charges, the user may have package preferences e.g., average rating resources of resource combination is maximized while total cost is minimized. It is not difficult to imagine many other applications of such queries. As another example consider a tourist interested in finding a number of tourist attractions at a destination where total distance traveling between them and total prices are minimized.

Since traditional querying models focus on finding list of items *rather than a list of item combinations*, the current strategy for arriving at such results is for users to review answer lists of items for possibly multiple queries and assemble and compare packages manually. In this article, we consider the idea of *skyline package queries over RDF data* which compute a pareto-optimal set of packages over RDF data, given multiple global or package preference criteria. There are three key fundamental challenges to be addressed: first is finding "relevant elementary or partial matches" i.e., items that meet *some* part of the description e.g., resources on just relational algebra, which can be combined with other partial matches to form complete results; second, is the number of joins required to stitch together the fine-grained (binary relations) representation of data in an RDF model to find matches; third is the computation of the skyline of combinations from elementary matches given package preferences. Depending on whether these tasks are done independently or holistically, query evaluation could be very expensive. The combination of these three challenges distinguish this from the few recent efforts on skyline [48] or top-k [23,44,45] package queries over single relations requiring no joins and in the latter case consider only a single preference criteria. While our previous work addresses the challenge of computing item skylines over RDF data models, the introduction of the packaging requirement in package queries limits the possibility of adapting the previously proposed approach in an efficient manner. Other work on skylining over joins in relational models [32,43,47] consider very restricted join patterns (single join [43], star-like schemas [47]) that are not applicable to RDF and also do not consider packages.

In this article, we present an efficient algebraic interpretation of package skyline queries over RDF in terms of a special operator and storage model. We also consider alternative interpretations that rely mostly on existing query operators and the mainstream vertical partitioned storage model for RDF. Specifically, we contribute the following: (1) a formalization of the logical query operator, *SkyPackage*, for package skyline queries over an RDF data model, (2) two families of query processing algorithms for physical operator implementation based on the traditional vertical partitioned storage model and a novel storage model here called *target descriptive, target qualifying* (TDTQ), and (3) an evaluation of the four algorithms proposed over synthetic and real world data. Specifically, we extend a previous paper [36] by proposing a more efficient algorithm, *SkyPackage*, for solving the skyline package problem and provide a more rigorous evaluation on the four algorithms. The rest of this article is organized as follows. The background and formalization of our problem is given in Sect. 2. Section 3 introduces the two algorithms based on the vertical partitioned storage model, and Sect. 4 presents an algorithm based on our TDTQ storage model. An empirical evaluation study is reported in Sect. 5, and related work is described in Sect. 6. The article is concluded in Sect. 7.

2   Background and Problem Definition

Consider a scenario where stores publish Semantic Web-enabled catalogs over which users can search based on products and some preferences. For example, a customer wishes to purchase milk, eggs, and bread from any combination of stores as long as the combination (*package*) offers an overall minimized total cost and overall maximized store ratings. Figure 1 shows an example ontology and data about stores and their ratings, items sold by each store and their prices. Here, one possible package is *aee* (i.e., buying milk from store *a*, eggs from store *e* and bread also from store *e*). Another possibility is *bee* (i.e., milk from store *b*, eggs and bread from store *e* as in the previous case). The bottom right of Fig. 1 shows the total price and average rating for packages *aee*, *bee* and *bge*. We see that *aee* is a better package than *bee* because it has a smaller total price and the same average rating. On the other hand, *bge* and *aee* are incomparable because although *bge*'s total price is worse than *aee*'s, its average rating is better.

Problem Definition. To contextualize the interpretation of such queries on RDF data, we build its algebraic interpretation on top of existing operators for interpretating SPARQL graph pattern queries. Observe that the example description contains some key components: a base graph pattern structure (the SPARQL queries in the figure). Graph pattern queries have return clauses (the

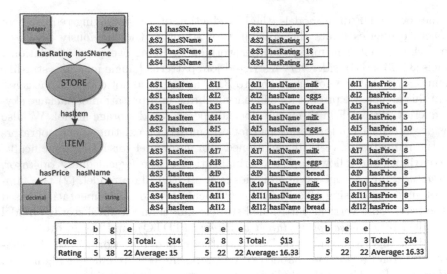

Fig. 1. Data For E-commerce example

SELECT clause) which denote the variables (*return variables*) whose bindings we are interested in including in the result. One of those variables, the (*target variable - ?store*) captures the *targets* of the query (*stores*) while the rest are part of the subquery structure that qualifies valid targets e.g., stores *should sell milk* and are called *target qualifing constraints - (?item, hasName, "milk")*. There are analogous graph patterns for qualifiers "selling *bread* and*eggs*". The second component of the query specification defines combinations or *packages* which can interpreted as a Fig. 2 of the results of above queries. The third component is pruning packages based on *preferences*,e.g., *minimizing the total price of package*. It is important to note that the preferences here are specified over aggregates of datatype properties (i.e., attributes) that are part of the description for targets (e.g., maximizing the average over store ratings) or target qualifiers, e.g., minimizing total price. The base graph patterns and crossproduct operations can be expressed using SPARQL's algebra or high-level language. The expression of preferences and item skyline queries are not supported by SPARQL although there have been some proposals for algebraic extensions to the SPARQL algebra. Our goal here, is to extend the SPARQL algebra to include operators that allow the algebraic expression of package skyline queries. We do not address SPARQL language grammar extensions here. The new operators to be introduced can be seen as a generalization the item skyline operators which when combined with operators for expressing graph pattern queries can express package skyline queries. In other words, in our framework, traditional item skyline queries will be viewed as package skyline queries where the package size is 1.

More formally, let D be a dataset with property relations $P_1, P_2, \ldots, P_m$ and GP be a graph pattern with triple patterns $TP_i, TP_j, \ldots, TP_k$ (TP_x means triple

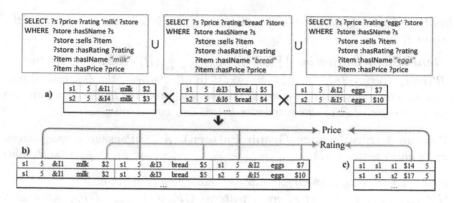

Fig. 2. Dataflow for the SkyPackage problem in terms of traditional query operators

pattern with property P_x). $[[GP]]_D$ denotes the answer relation for GP over D, i.e., $[[GP]]_D = P_i \bowtie P_j \bowtie \ldots \bowtie P_k$. Let $var(TP_x)$ and $var(GP)$ denote the variables in the triple and graph pattern respectively and $r \in var(GP)$ denote the target variable. Note that the rest of the graph pattern is used to define the target i.e target description and relationship between qualifiers and targets.

For preference specification, we begin by reviewing the formalization for preferences given in [32]. Let $Dom(a_1), \ldots, Dom(a_d)$ be the domains for the columns $a_1, \ldots, a_d$ in a d-dimensional tuple $t \in [[GP]]_D$. Given a set of attributes $B \subseteq A'$, a preference PF over the set of tuples $[[GP]]_D$ is then defined as $PF := (B; \prec_{PF})$, where $\prec_{PF}$ is a strict partial order on the domain of B. Given a set of preferences $PF_1, \ldots, PF_m$, their combined Pareto preference PF is defined as a set of equally important preferences.

For a set of d-dimensional tuples R and preference $P = (B; \prec_P)$ over R, a tuple $r_i \in R$ dominates tuple $r_j \in R$ based on the preference P (denoted as $r_i \prec_P r_j$), iff $(\forall(a_k \in B)(r_i[a_k] \preceq r_j[a_k]) \wedge \exists(a_l \in B)(r_i[a_l] \prec r_j[a_l]))$.

Definition 1 (Skyline Query). *To adapt preferences to graph patterns, we associate a preference with the property (assumed to be a datatype property) on whose object the preference is defined or the preference property e.g., the property* rating. *Let PF_i denote a preference with preference property as P_i. Then, for a graph pattern $GP = TP_1, \ldots, TP_m$ and a set of preferences $PF = PF_i, PF_j, \ldots, PF_k$, a skyline query $SKYLINE[[[GP]]_D, PF]$ returns the set of tuples from the answer of GP such that no other tuples dominate them with respect to PF and they do not dominate each other. The extension of the skyline operator to packages is based on two functions* Map *and* Generalized Projection.

Definition 2. *Let $\mathcal{F} = \{f_1, f_2, \ldots, f_k\}$ be a set of k mapping functions such that each function $f_j(B)$ takes a subset of attributes $B \subseteq A$ of a tuple t, and returns a value x.*

Map $\hat{\mu}_{[\mathcal{F}, \mathcal{X}]}$ *(adapted from [32]) applies a set of k mapping functions $\mathcal{F}$ and transforms each d-dimensional tuple t into a k-dimensional output tuple t'*

defined by the set of attributes $\mathcal{X} = \{x_1, x_2, \ldots, x_k\}$ with x_i generated by the function f_i in $\mathcal{F}$.

Generalized Projection $\prod_{colr_x, colr_y, colr_z, \hat{\mu}_{[\mathcal{F}, \mathcal{X}]}} (R)$ *returns the relation* $R'(colr_x, colr_y, colr_z, \ldots, x_1, x_2, \ldots, x_k)$. *In other words, the generalized projection outputs a relation that appends the columns produced by the map function to the projection of R on the attributes listed, i.e., $colr_x, colr_y, colr_z$.*

Definition 3 (SkyPackage Graph Pattern). *A SkyPackage graph pattern query is graph pattern $GP_{[r, \{c_1, c_2, \ldots, c_N\}, \mathcal{F} = \{f_1, f_2, \ldots, f_k\}, \{PF_i, PF_j, \ldots, PF_k\}]}$ such that:*

1. c_i *is a triple pattern specifying a target qualifying constraint.*
2. $GP_{\{c_1, c_2, \ldots, c_N\}}$ *is the set of graph patterns $GP_{c_1}, GP_{c_2}, \ldots, GP_{c_N}$ associated with target qualifying constraints $c_1, \ldots, c_N$ e.g., each of the three graph patterns in Fig. 1 represent a GP_{c_i}.*
3. r *is a set of return variables such that $r_i \in r$ implies $r_i \in \mathrm{var}(GP_i)$ is called the target of the query, e.g., stores.*
4. PF_i *is the preference specified on the property P_i whose mapping function is f_i or more specifically, $PF_i = (f_i(P_i); \prec_{PF_i})$.*

The answer to a SkyPackage graph pattern query R_{SKY} can be described algebraically as the result of a sequence of operators in the following manner:

1. $R_{product} = [[GP_{c_1}]] \times [[GP_{c_2}]] \times \cdots \times [[GP_{c_N}]]$ *such that $[[GP_{c_x}]]$ is the result of evaluating the branch of the union query with constraint c_x. Figure 2b shows the partial result of the crossproduct of the three subqueries in (a) based on the 3 constraints on milk, bread and eggs.*
2. $R_{project} = \prod_{r_1, r_2, \ldots r_N, \hat{\mu}_{[\mathcal{F} = \{f_1, f_2, \ldots f_k\}, \mathcal{X} = \{x_1, x_2, \ldots x_k\}]}} (R_{product})$ *where*

   r_i *is the column for the target variable in subquery i's result, $f_1 : (dom_{c_1}(o_1) \times dom_{c_2}(o_1) \times \cdots \times dom_{c_N}(o_1)) \to \mathbb{R}$ where $dom_{c_1}(o_1)$ is the domain of values for the column representing the object of P_1, e.g., column for object of hasPrice, in $[[GP_{c_1}]]$. The functions in our example would be $total_{hasPrice}$, $average_{hasRating}$. The output of this step is shown in Fig. 2c.*
3. $SKYLINE[R_{project}, \{PF_{P'_i}, PF_{P'_j}, \ldots, PF_{P'_k}\}]$ *such that $PF_{P'_i}$ is the preference defined on the aggregated columns produced by the map function (denoted by P'_i), e.g., minimizing total price.*

Our example in this model is $GP_{[r, X, \mathcal{F}, PF]}$, where

- $r = ?store$
- $X = \{(?item, hasName, "milk"), (?item, hasName, "bread"),$ $(?item, hasName, "eggs")\}$
- $\mathcal{F} = \{SP = sum()_{price}, SR = average()_{rating}\}$
- $PF = \{(SP; \prec_{min}), (SR; \prec_{max})\}$

2.1   Related Work

Although much research has been done in the area of traditional skyline queries, package skyline queries have not received a generous amount of attention. Our contribution is unique from previous work in that we provide algorithms whose (package) skyline result contains elements of cardinality greater than one. In this section, we present previous work related to our study of skyline packages. We begin with a historical perspective of how the skyline query came about and an overview of some of the earlier solutions proposed outside of a database context. Then we provide an overview of solutions within a database context and their correspondence to single-relations, multi-relations, and composite top-k queries.

The skyline query problem originally arose in the theory field in the 1960s, and the skyline set was coined as the *Pareto set*. This problem became known as the *maximal vector problem* [28,34], whose solution (e.g., skyline) is called *maximal vectors* [6] or *admissible points* [2], and is similar to the *contour problem* [31] and *convex hull problem*. Solutions to the maximal vector problem were proposed in [5,6,28]; however, these solutions cannot scale to large databases because they require all data to be in memory.

Single-relation Skyline Algorithms. The first proposed method of applying the maximal vector problem to databases was [10] and the term *skyline queries* was coined. Since then, the skyline query problem has often been referred to as a secondary/external storage version of the maximal vector problem [24]. [10] originally introduced and provided a block nested loops, divide-and-conquer, and B-tree-based algorithms. Later, [18] introduced a sort-filter-skyline algorithm that is based on the same intuition as BNL, but uses a monotone sorting function to allow for early termination. Unlike [10,18,41], which has to read the whole database at least once, index-based algorithms [26,33] allow one to access only a part.

Multi-relation Skyline Algorithms. All of the previous algorithms are designed to work on a single relation. As the Semantic Web matures and RDF data is populated, there has been an increase in research involving multi-relational skyline queries. When queries involve aggregations, multiple relations must be joined before any of the above techniques can be used. Implicitly, the first work that deals with the problem of skyline over multiple relations via joins is [27]. Given a query that joins two relations and filters the result using a WHERE clause, the authors propose a method to overcome empty results known as *query relaxation*, which relaxes the join selection thus making the query more flexible. Unlike our work, they do not focus on preference queries.

Vlachou et al. [43] proposed a novel algorithm called SFSJ (sort-first skyline-join) that computes the complete skyline. Given two relations, access to the two relations are alternated using a pulling strategy, known as adaptive pulling, that prioritizes each relation based on the number of mismatched join values. SFSJ takes advantage of its early termination condition, which gives rise to its performance, when the two regions from each relation meet a certain condition,

Although the algorithm has no limitations on the number of skyline attributes, it is limited by two relations.

Recently, [16] introduced three skyline algorithms that are based on the concept of a *header point*, which allows some nonskyline tuples to be discarded before proceeding to the skyline processing phase. Raghavan and Rundensteiner [32] introduced a sky-join operator that gives the join phase a small knowledge of a skyline. An index-based, non-join skyline algorithm was proposed in [48]. Khabbaz and Lakshmanan [23] proposes a framework for collaborative filtering using a variation of top-k. However, their set of results do not contain packages but single items.

Composite Top-k Algorithms. Up until now, very little research has been conducted in the area of package, or composite, queries. Such previous work mostly aims at providing composite results in a recommendation system [44,45]. Xie [45] uses top-k techniques to provide a composite recommendation for travel planning. Since finding packages is complex and time consuming, most have oriented their work towards *approximating* the desired packages [44]. Unlike our goal, which is to provide the user with *all* correct results, this approach limits the user from seeing the complete results. Top-k is useful when ranking objects is desired. However, top-k is prone to discard tuples that have a 'bad' value in one of the dimensions, whereas a skyline algorithm will include this object if it is not dominated.

3   Algorithms for Package Skyline Queries over Vertical Partitioned Tables

We present in this section two approaches: *Join, Cartesian Product, Skyline* (*JCPS*) and *RDFSkyJoinWithFullHeader-Cartesian Product, Skyline* (*RSJFH-CPS*), for solving the package skyline problem. These approaches assume data is stored in vertically partitioned tables (VPTs) [1].

3.1   *JCPS* Algorithm

The formulation of the package skyline problem suggests a relatively straightforward algorithm involving multiple *joins*, followed by a *Cartesian product* to produce all combinations, followed by a single-table *skyline* algorithm (e.g., block-nested loop), called *JCPS*.

Consider the VPTs *hasIName*, *hasSName*, *hasItem*, *hasPrice*, and *hasRating* obtained from Fig. 1. Solving the skyline package problem using *JCPS* involves the following steps. First, we join all tables and perform a Cartesian product twice on I to obtain all store packages of size 3, as shown in Fig. 3a. As the product is being computed, the *price* and *rating* attributes are aggregated, as shown in Fig. 3b. Afterwards, a single-table skyline algorithm is performed to discard all dominated packages with respect to total price and average rating.

Algorithm 1 contains the pseudocode for such an algorithm. Solving the skyline package problem using *JCPS* requires all VPTs to be joined together (line 2), denoted as I. To obtain all possible combinations (i.e., packages) of targets, multiple specialized Cartesian products are performed on I (lines 3–6). A modified Cartesian product, denoted as $\otimes$, is implemented as a subroutine to ensure no duplicate target constraints (e.g., milk) are inlcuded. Afterwards, *equivalent* skyline attributes are aggregated (lines 7–10). Equivalent skyline attributes, for example, of the e-commerce motivating example would be price and rating attributes. Aggregation for the price of milk, eggs, and bread would be performed to obtain a total price. Finally, line 11 applies a single-table skyline algorithm to remove all dominated packages.

Algorithm 1: JCPS

Input: $VPT_1, VPT_2, \ldots VPT_x$ containing skyline attributes $s_1, s_2, \ldots, s_y$, and an aggregation function $\mathcal{A}(T)$ on some table T

Output: Package Skyline $\mathcal{P}$

1: $n \leftarrow$ package size
2: $I \leftarrow VPT_1 \bowtie VPT_2 \bowtie \cdots \bowtie VPT_x$
3: $I_2 \leftarrow I \times I$
4: **for all** $i \in [1, n-2]$ **do**
5:     $I_2 \leftarrow I_2 \otimes_{i,col} I$
6: **end for**
7: $M_1 \leftarrow \mathcal{A}(I_{2_{s_1}})$
8: **for all** $i \in [2, y]$ **do**
9:     $M_i \leftarrow \mathcal{A}(M_{(i-1)_{s_i}})$
10: **end for**
11: $\mathcal{P} \leftarrow$ skyline(M_y)
12: **return** $\mathcal{P}$

subroutine $\otimes(T_1, T_2, col, iteration)$

1: $list \leftarrow \varnothing$
2: **for all** $t1 \in T_1$ **do**
3:     **for all** $t2 \in T_2$ **do**
4:        $list$.add$(T_2(col))$
5:        **for all** $i \in [1, iteration]$ **do**
6:           **if** ($!list$.contains$(T_1(iteration + col))$) **then**
7:              $list$.add$(T_1(iteration + col))$
8:           **else**
9:              $list \leftarrow \varnothing$
10:              break
11:           **end if**
12:           $T_3 \leftarrow T_1(col) \bowtie T_2(col)$
13:        **end for**
14:        $list \leftarrow \varnothing$
15:     **end for**
16: **end for**
17: **return** T_3

The limitations of such an algorithm are fairly obvious. First, many unnecessary joins are performed. Furthermore, if the result of joins is large, the input to the Cartesian product operation will be very large even though it is likely

item	price	store	rating
milk	2	A	5
eggs	7	A	5
⋮			

⊗

item	price	store	rating
milk	2	A	5
eggs	7	A	5
⋮			

⊗

item	price	store	rating
milk	2	A	5
eggs	7	A	5
⋮			

(a) Cartesian Product of Join Result

item1	item2	item3	store1	store2	store3	total price	average rating
milk	eggs	bread	A	B	C	10	5
milk	eggs	bread	B	B	A	7	5
⋮							

(b) Cartesian Product Result

Fig. 3. Resulting tables of $JCPS$

that only a small fraction of the combinations produced will be relevant to the skyline. The exponential increase of tuples after the Cartesian product phase will result in a large number of tuple-pair comparisons while performing a skyline algorithm. In addition, duplicates will have to be eliminated. To gain better performance, it is crucial that some tuples be pruned before entering into the Cartesian product phase, which is discussed next.

3.2 *RSJFH-CPS* Algorithm

A pruning strategy that prunes the input size of the Cartesian product operation is crucial to achieving efficiency. One possibility is to exploit the following observation: *skyline packages can be made up of only target resources that are in the skyline result when only one constraint (e.g., milk) is considered* (note that a query with only one constraint is equivalent to an item skyline query).

Lemma 1. *Let $\rho = \{p_1 p_2 \ldots p_n\}$ be a package of size n (i.e., containing n target resources), $\mathcal{P}$ be the set of all skyline packages, and $p_1', p_2', \ldots, p_n'$ be other target resources with respect to a qualifying constraint $C_1, C_2, \ldots C_n$. If $\rho \in \mathcal{P}$, then $p_m \preceq_{C_m} p_m'$ for all $1 \leq m \leq n$.*

Proof. Let $\rho' = \{p_1 p_2 \ldots p_n'\}$, where $p_n \preceq_{C_n} p_n'$, and let $x_1, x_2, \ldots, x_m$ be the preference attributes for p_n and p_n'. Since $p_n \preceq_{C_n} p_n'$, $p_n[x_j] \preceq p_n'[x_j]$ for some $1 \leq j \leq n$. Therefore, $\mathcal{A}_{1 \leq i \leq n}(p_i[x_j]) \preceq \mathcal{A}_{1 \leq i \leq n}(p_i'[x_j])$, where $\mathcal{A}$ is a monotonic aggregation function and $p_i' \in \rho'$. Since for any $1 \leq k \leq n$, where $k \neq j$, $\mathcal{A}_{1 \leq i \leq n}(p_i[x_k]) = \mathcal{A}_{1 \leq i \leq n}(p_i'[x_k])$. This implies that $\rho \preceq \rho'$. Thus, ρ' is not a skyline package. □

As an example, let $\rho = \{p_1 p_2\}$ and $\rho' = \{p_1 p_2'\}$ and x_1, x_2 be the preference attributes for p_1, p_2, p_2'. We define the attribute values as follows: $p_1 = (3, 4), p_2 = (3, 5)$, and $p_2' = (4, 5)$. Assuming the lowest values are preferred, $p_2 \preceq p_2'$ and

$p_2[x_1] \preceq p'_2[x_1]$. Therefore, $\mathcal{A}_{1 \leq i \leq 2}(p_i[x_1]) \preceq \mathcal{A}_{1 \leq i \leq 2}(p'_i[x_1])$. In other words, $(p_1[x_1] + p_2[x_1]) \preceq (p_1[x_1] + p'_2[x_1])$, i.e., $(3 + 3 = 6 \preceq 7 = 3 + 4)$. Since all attribute values except $p'_2[x_1]$ remained unchanged, by definition of skyline we conclude $\rho \preceq \rho'$.

This lemma suggests that the skyline phase can be pushed ahead of the Cartesian product step as a way to prune the input of the $JCPS$. Even greater performance can be obtained by using a skyline-over-join algorithm, $RSJFH$ [16], that combines the skyline and join phase together. $RSJFH$ takes as input two VPTs sorted on the skyline attributes. We call this algorithm $RSJFH$-CPS. This lemma suggests that skylining can be done in a divide-and-conquer manner where a skyline phase is introduced for each constraint, e.g., milk, (requiring 3 phases for our example) to find all potential members of skyline packages which may then be fed to the Cartesian product operation.

Given the VPTs $hasIName$, $hasSName$, $hasItem$, $hasPrice$, and $hasRating$ obtained from Fig. 1, solving the skyline package problem using $RSJFH$-CPS involves the following steps:

1. $I^2 \leftarrow hasSName \bowtie hasItem \bowtie hasRating$
2. For each target t (e.g., milk)
   (a) $I^1_t \leftarrow \sigma_t(hasIName) \bowtie hasPrice$
   (b) $S_t \leftarrow RSJFH(I^1_t, I^2)$
3. Perform a Cartesian product on all tables resulting from step 2b
4. Aggregate the necessary attributes (e.g., price and rating)
5. Perform a single-table skyline algorithm

Figure 4a shows two tables, where the left one, for example, depicts step (a) for milk, and the right table represents I^2 from step 1. These two tables are sent as input to $RSJFH$, which outputs the table in Fig. 4b. These steps are done for each target, and so in our example, we have to repeat the steps for *eggs* and *bread*. After steps 1 and 2 are completed (yielding three tables, e.g., *milk*, *eggs*, and *bread*), a Cartesian product is performed on these tables, as shown in Fig. 4c, which produces a table similar to the one in Fig. 3b. Finally, a single-table skyline algorithm is performed to discard all dominated packages.

Algorithm 2 shows the pseudocode for $RSJFH$-CPS The main difference between $JCPS$ and $RSJFH$-CPS appears in line 5–8. For each target, a *select* operation is done to obtain all like targets, which is then joined with another VPT containing a skyline attribute of the targets. This step produces a table for *each* target. After the remaining tables are joined, denoted as I^2 (line 4), each target table I^1_i along with I^2 is sent as input to $RSJFH$ for a skyline-over-join operation. The resulting target tables undergo a Cartesian product phase (line 9) to produce all possible combinations, and then all equivalent attributes are aggregated (lines 10–12). Lastly, a single-table skyline algorithm is performed to discard non-skyline packages (line 13). Since a skyline phase is introduced early in the algorithm, the input size of the Cartesian product phase is decreased, which significantly improves execution time compared to $JCPS$.

hasIName ⋈ hasPrice				
&I1	milk	&I1	2
&I4	milk	&I4	3

⋈

hasSName ⋈ hasItem ⋈ hasRating					
&S1	A	&S1	&I1	&S1	5
&S1	A	&S1	&I2	&S1	5

(a) *RSJFH* (skyline-over-join) for milk

item	price	store	rating
milk	2	A	5
milk	3	B	5
⋮			

(b) *RSJFH*'s result for milk

item	price	store	rating
milk	2	A	5
⋮			

✖

item	price	store	rating
eggs	3	B	5
⋮			

✖

item	price	store	rating
bread	5	A	4
⋮			

(c) Cartesian product on all targets (e.g., milk, eggs, and bread)

Fig. 4. Resulting tables of *RSJFH*

Algorithm 2: RSJFH-CPS

Input: $VPT_1, VPT_2, \ldots VPT_x$ containing skyline attributes $s_1, s_2, \ldots, s_y$, and corresponding aggregation functions $\mathcal{A}_{s_1}(T), \mathcal{A}_{s_2}(T), \ldots, \mathcal{A}_{s_y}(T)$ on table T

Output: Package Skyline $\mathcal{P}$

1: $n \leftarrow$ package size
2: $t_1, t_2, \ldots, t_n \leftarrow$ targets of the package
3: VPT_1 contains targets and VPT_2 contains a skyline attribute of the targets
4: $I^2 \leftarrow VPT_3 \bowtie \cdots \bowtie VPT_x$
5: **for all** $i \in [1, n]$ **do**
6:    $I_i^1 \leftarrow \sigma_{t_i}(VPT_1) \bowtie VPT_2$
7:    $S_i \leftarrow RSJFH(I_i^1, I^2)$
8: **end for**
9: $T \leftarrow S_1 \times S_2 \times \cdots \times S_n$
10: $M_1 \leftarrow \mathcal{A}(T_{s_1})$
11: **for all** $i \in [2, y]$ **do**
12:    $M_i \leftarrow \mathcal{A}(M_{(i-1)_{s_i}})$
13: **end for**
14: $\mathcal{P} \leftarrow \text{skyline}(M_y)$
15: **return** $\mathcal{P}$

MILK		EGGS		BREAD		RATING	
store	price	store	price	store	price	store	value
a	2	g	5	e	3	e	5
b	3	a	7	b	4	i	5
f	3	c	8	a	5	c	12
e	4	e	8	i	8	f	13
g	6	h	9	g	5	h	14
e	9	b	10	f	6	g	18
h	9	d	10	h	6	a	20
d	10			d	10	d	21
						b	22

Fig. 5. Target qualifying (*milk, eggs, bread*) and target descriptive (*rating*) tables for E-commerce example

4 Algorithms for Package Skyline Queries over the TDTQ Storage Model

4.1 The TDTQ Storage Model

While the previous two approaches, *JCPS* and *RSJFH-CPS*, rely on VPTs, the next approach is a multistage approach in which the first phase is analogous to the build phase of a hash-join. In our approach, we construct two types of tables: *target qualifying* tables and *target descriptive* tables, called *TDTQ*. Target qualifying tables are constructed from the target qualifying triple patterns (*?item hasIName "milk"*) and the triple patterns that associate them with the targets (*?store sells ?item*). In addition to these two types of triple patterns, a triple pattern that describes the target qualifier that is associated with a preference is also used to derive the target qualifying table. In summary, these three types of triple patterns are joined per given constraint and a table with the target and preference attribute columns are produced. The left three tables in Fig. 5 show the the target qualifying tables for our example (one for each constraint). The target descriptive tables are constructed per target attribute that is associated with a preference, in our example *rating* for stores. These tables are constructed by joining the triple patterns linking the required attributes and produce a combination of attributes and preference attributes (store name and store rating produced by joining hasRating and hasSName). The rightmost table in Fig. 5 shows the target descriptive table for our example.

We begin by giving some notations that will aid in understanding of the TDTQ storage model. In general, the build phase produces a set of partitioned tables $T_1, \ldots, T_n, T_{n+1}, \ldots, T_m$, where each table T_i consists of two attributes, denoted by T_i^1 and T_i^2. We omit the subscript if the context is understood or if the identification of the table is irrelevant. $T_1, \ldots, T_n$ are the target qualifying tables where n is the number of qualifying constraints. $T_{n+1}, \ldots, T_m$ are the target descriptive tables, where $m - (n + 1) + 1 = m - n$ is the number of target attributes involved in the preference conditions.

4.2 *CPJS* and *SkyJCPS* Algorithms

Given the TDTQ storage model presented previously, one option for computing the package skyline would be to perform a Cartesian product on the target qualifying tables, and then joining the result with the target descriptive tables. We call this approach *CPJS* (*Cartesian product, Join, Skyline*), which results in exponential time and space complexity. Given n targets and m target qualifiers, n^m possible combinations exist as an intermediate result prior to performing a skyline algorithm. Each of these combinations is needed since we are looking for a set of packages rather than a set of points. Depending on the preferences given, additional computations such as aggregations are required to be computed at query time. Our objective is to find all package skylines *efficiently* by eliminating unwanted tuples before we perform a Cartesian product. Algorithm 3 shows the *CPJS* algorithm for determining package skylines.

Algorithm 3: CPJS

Input: $T_1, T_2, \ldots T_n, T_{n+1}, \ldots, T_m$
Output: Package Skyline $\mathcal{P}$
1: $I \leftarrow T_1 \times T_2 \times \cdots \times T_n$
2: **for all** $i \in [n+1, m]$ **do**
3:    $I \leftarrow I \bowtie T_i$
4: **end for**
5: $\mathcal{P} \leftarrow \text{skyline}(I)$
6: **return** $\mathcal{P}$

CPJS begins by finding all combinations of targets by performing a Cartesian product on the target qualifying tables (line 1). This resulting table is then joined with each target descriptive table, yielding a single table (line 3). Finally, a single-table skyline algorithm is performed to eliminate dominated packages (line 5).

Applying this approach to the data in Fig. 5, one would have to compute all 448 possible combinations before performing a skyline algorithm. The number of combinations produced from the Cartesian product phase can be reduced by initially introducing a skyline phase on each target, e.g., milk, as we did in for *RSJFH-CPS*. We call this algorithm *SkyJCPS*. Although similarities to *RSJFH-CPS* can be observed, *SkyJCPS* yields better performance due to the reduced number of joins. Figure 4a clearly illustrates that *RSJFH-CPS* requires four joins before an initial skyline algorithm can be performed. All but one of these joins can be eliminated by using the TDTQ storage model. To illustrate *SkyJCPS*, given the TDTQ tables in Fig. 5, solving the skyline package problem involves the following steps:

1. For each target qualifying table TQ_i (e.g., milk)
   (a) $I_{TQ_i} \leftarrow (TQ_i) \bowtie rating$
   (b) $I'_{TQ_i} \leftarrow \text{skyline}(I_{TQ_i})$
2. $CPJS(I'_{TQ_1}, \ldots, I'_{TQ_i}, rating)$

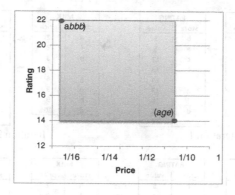

Fig. 6. Skyline region

Since the dominating cost of answering skyline package queries is the Cartesian product phase, the input size of the Cartesian product can be reduced by performing a single-table skyline algorithm over each target.

4.3 SkyPackage Algorithm

Our algorithm attempts to discern the skyline set by stepping through sorted sub-lists in a way that guarantees we move towards the skyline set. This strategy relies on being able to discover the best package in one dimension, which means we are guaranteed that no future packages will be dominated by this one. From Fig. 5, the package consisting of the first tuple from each of the target qualifying tables (age) constitutes a package skyline. Also, the package that has the best rating can be found by looking at the rating table for the highest rating, which is b. Thus, package (bbb) is a package skyline. To illustrate this, consider the two packages in Fig. 6, {(age),$10,14} and {(bbb), $17,22}. All future package skylines must fall between these two packages. We depict this in the shaded area. Any package that does not lie within this region can be immediately discarded. However, *candidate* packages may fall within this area and will need to be checked for membership in the skyline package.

Pruning and Early Termination. Since performing a Cartesian product is expensive and its output size to the number of package skylines ratio is very large, it is desirable to decrease this intermediate result. Therefore, we have to eliminate targets that cannot possibly be in the skyline set when packaged with any other available targets. While the naive approach would have to compose the packages and then perform a skyline to filter out unwanted tuples, our pruning technique offers *local prunability* that prunes tuples, i.e., targets, from individual target qualifiers before we form the packages.

To illustrate the concept of local pruning, consider the *milk* and *rating* tables in Fig. 5. We present in Fig. 7 four iterations where each iteration indicates a new tuple being examined. As we look at each store in the *milk* table, we probe

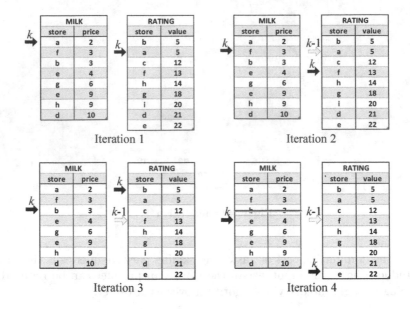

Fig. 7. Pruning example

the *rating* table and keep a pointer at its value. The first tuple examined, i.e., first iteration, in the *milk* table is $(a, 2)$. We probe the *rating* table to locate store a, and keep a pointer there for the next iteration(s). In the second iteration, we examine the next tuple $(f, 3)$ and probe the *rating* table again. We compare its value against the previous pointer, denoted as $k-1$. Since the current store is better than the previous store, we remove the pointer from store a and continue to the third iteration. As usual, the current tuple $(b, 3)$ is used to probe the *rating* table. In this case, the current pointer k $(b, 5)$ is worse than the previous pointer $k - 1$ $(f, 13)$, and thus we prune the tuple $(b, 3)$. Because the current pointer that points to $(f, 13)$ is no better than the previous pointer, we save this pointer and examine the next tuple as shown in iteration 4.

The concept of local prunability is formalized in the following lemma.

Lemma 2 (Pruning). *Let $T_j[k]$ be the value of object k in table T_j, then $\forall k \in T_i{}^1, i \in [1, n]$, if $\exists j \in [n+1, m]$ such that $T_j[k-1] \prec T_j[k]$, and $T_i[k-1] \preceq T_i[k]$ then object k does not produce a package skyline and can be pruned from T_i.*

Proof. Since T_i is sorted in the preprocessing phase of our database, we know $T_i[k - 1] \preceq T_i[k]$. Assume that k produces a package skyline (is part of the combination). Then, if $T_j[k] \preceq T_j[k-1]$, object $k-1$ dominates k, thus k cannot be part of the package skyline. $\square$

Lemma 2 ensures that any combination where k appears is not a package skyline. If we denote the size of each table T as $|T|$, then for each tuple pruned in T_i, the size of the resulting Cartesian product is reduce to $|T_1| \times |T_2| \times \cdots \times$

MILK			EGGS			BREAD	
store	price		store	price		store	price
a	2		g	5		e	3
b	3		a	7		b	4
			b	10			

Fig. 8. Pruning and early termination result

package	price	rating
age	10	14.3
agb	11	20
bgb	12	20.7
bab	14	21.3
bbb	17	22

Fig. 9. Skyline package result

$(|T_i| - 1) \times \cdots \times |T_n|$. To illustrate this, after tuple $(b, 3)$ is pruned in Fig. 4, the Cartesian product size is reduced from 448 to 392 tuples.

We define the *previous object* to be a pointer to the best last seen object and the pointer is updated when a better object is examined. In Lemma 2, we denote the current object as k and the previous object as $k - 1$. If Lemma 2 is not satisfied, the pointer that once pointed to $k - 1$ is updated to point to k.

To increase performance of our algorithm, we utilize the following early termination strategy for each resource.

Lemma 3 (Early Termination). *If k is the current object in T_i and for all $j \in [n + 1, m], T_j[k]$ is the best in T_j, then stop examining T_i and continue to $T_{i+1}, i + 1 \leq n$.*

Proof. Assume there exists an object $k + 1$ in T_i that has not been examined. Then $T_i[k] \preceq T_i[k + 1]$, and $T_j[k] \preceq T_j[k + 1]$. If $T_j[k + 1] \neq T_j[k]$, then $T_j[k] \prec T_j[k + 1]$. Then for any object after k, $T_i[k] \preceq T_i[k+1] \preceq T_i[k+2] \cdots$. Therefore, every object after k is dominated.   □

Lemma 3 allows us to stop examining tuples in a given table when the best object is seen in the target descriptive tables , i.e., *rating* table. For example, in Fig. 7, since b has the highest rating, we stop scanning the *milk* table once b is examined and prune all tuples below it. If a target qualifying table does not contain the best object from the target descriptive table, we choose the next best object such that the target qualifying table contains this object.

After pruning, our next phase is performing a Cartesian product. As the product is produced, if there exist any tuples that do not satisfy (local) hard constraints, we discard these. Figure 8 shows the resulting tables after pruning.

After the pruning phase is complete, a Cartesian product is performed among the target qualifying tables and joined with the target descriptive table(s) for aggregation. In Fig. 8, a Cartesian product involving the *milk*, *eggs*, and *bread* tables is performed to find (1) all packages and (2) total price. The intermediate result is then joined with the *rating* table to find the average rating. If there exists any tuples that do not satisfy (global) hard constraints, we discard these. A skyline algorithm is performed to remove any packages that are not skyline packages. The final skyline package set is shown in Fig. 9.

Discussion. Now that we have provided a concrete example of the SkyPackage algorithm, we will now explain the pseudocode in Algorithm 4. Lines 1–3 of the algorithm explain some notations that are used within the algorithm. Once the query is issued, we examine each of the n tables (line 4) one row at time (line 6), keeping a pointer p that points to the $n + 1$ table that has the best value. With each iteration, we initialize *ptr* (line 5) to the first object in T_i mapped to T_{n+1}. Then we check whether Lemma 2 holds (line 7). Lines 8–14 handles the case when two consecutive objects have the same value in T_i. In this case, the tuple with the worse value in T_{n+1} is pruned. Lines 15–17 are similar to lines 8–14 except the equality checks are done on T_{n+1} rather than T_i. That is if two objects have the same value in T_{n+1}, we prune the tuple that has the smallest value in T_i. In line 18, we reach our early termination check, Lemma 3. We can safely stop examining the current table when we access an object that has the lowest value in t_{n+1}. It can easily be showed that any tuple after this one cannot be in the skyline set. At this point, *ptr* can no longer be updated since any subsequent tuple will have a higher value in t_{n+1}. If local constraints are given, we perform a check in line 19 to determine whether the current tuple satisfies the constraints. If the current tuple is not satisfied, all tuples below and including this one are pruned. We then join the tables, removing any tuples that do not satisfy any global constraints. Lastly, any known skyline algorithm is performed.

5   Sesame Integration Framework

5.1   Sesame

Sesame [14] is an open-source RDF database implemented in Java whose architecture allows for persistent storage of RDF data and querying of that data. We chose Sesame as our RDF engine for a number of reasons. First, since Sesame is a server-based application, it allowed us to store and query data on the Semantic Web remotely. Second, Sesame does not require a specific communication protocol or storage mechanism to be used.

5.2   Framework

The data that was used in the framework was the MovieLens[1] dataset, which was converted to RDF format using the Jena API [15]. Its ontology is depicted in Fig. 10. We used a server with Linux and an Apache Tomcat Web container.

---

[1] http://www.grouplens.org/node/73/

Algorithm 4: SkyPackage

Input: $T_1, T_2, \ldots T_n, T_{n+1}, \ldots, T_m$
Output: Package Skyline $\mathcal{P}$
1: $v_k(i) \leftarrow$ the value of object k in table i
2: $k \leftarrow$ the current object (i.e., row)
3: $ptr \leftarrow$ the best object
4: **for all** $i \in [1, n]$ **do**
5: $ptr \leftarrow v_x(n+1)$, $x \leftarrow$ first tuple in i
6: **for all** $k \in t_i$ **do**
7:   check whether Lemma 2 holds
8:   **if** $v_k(i) = v_{k-1}(i)$ **then**
9:    **if** $v_k(n+1) > v_{k-1}(n+1)$ **then**
10:     prune $(k, v_k(i))$
11:    **else**
12:     prune $(k-1, v_{k-1}(i))$
13:    **end if**
14:   **end if**
15:   **if** $v_k(n+1) = v_{k-1}(n+1)$ **then**
16:    prune max $\{(k, v_k(i)), (k-1, v_{k-1}(i))\}$
17:   **end if**
18:   check whether Lemma 3 holds
19:   check k against local constraints
20: **end for**
21: **end for**
22: cross product, remove tuples not satisfying global constraints
23: skyline

A Web-based interface was designed to allow users to query a subset of the MovieLens dataset. Although any package-related query can be supported, for the purpose of this article, we chose to support a query of the following form.

Query 1. *Given n movie-raters, find packages of n movies such that the average rating of all the movies is high and each movie-rater has rated at least one of the movies.*

When the user provides preferences using the Web-based interface, the information is sent to the SPARQL adapter that dynamically creates a SPARQL query. After the SPARQL query is executed and results of this query is available, the *SkyPackage* algorithm finds and presents the skyline package(s) that meet the user's preferences.

Data Storage. One option of storing RDF triples is to store them in a text file. However, this is inefficient for large numbers of triples and a solution involving indexing (e.g., database management system) is more appropriate. Relational databases, such as MySQL and Oracle, can be used to store such data, but are usually not optimized for such. Databases that are optimized for storage of RDF triples are called *triplestores*.

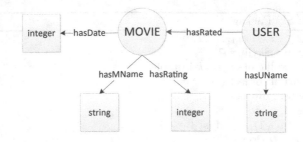

Fig. 10. MovieLens ontology

Sesame triplestore stores RDF triples in a *repository*. Sesame abstracts from any particular storage mechanism allowing a variety of repositories to be handled, including RDF triplestores and relational databases. Sesame offers several repositories in which to store data and all differs in where the data is stored. Two popular repositories are *memory* store and *native* store, corresponding to storage in-memory and on-disk, respectively. We used the native store configuration in our framework since it offers a better scalability solution for larger data sets as it is not limited to size of available memory. For native store, Sesame provides B-tree indexes on any combination of the subjects, properties, and objects. The index key(s) consist of subject (s), predicate (p), object (o), and context (c). The order in which these fields are listed determines the usability of the index. We chose to have as the index keys: *spoc* and *opsc*.

Data Retrieval. In order to retrieve data from Sesame's repository, we devised a skyline package operator, whose ultimate goal is to form queries that when executed will construct the TQ and TD tables. A description is given on how the queries for the TQ tables are constructed, followed by a similar description on how to construct the TD table.

We define *SP(C, A, PF)* as the skyline package operator that takes as arguments a list of constraints C, a list of properties A, and a list of preferences PF. In addition, we assume that the following VPTs exist: $vpt_1, vpt_2, \ldots, vpt_m,$ $vpt_{m+1}, \ldots, vpt_n$. Moreover, for the purpose of illustration, we assume $vpt_1, \ldots,$ vpt_m and $vpt_{m+1}, \ldots vpt_n$ are sufficient to construct the TQ and TD tables, respectively. The arguments for the SP operator are defined as follows:

- $C = (c_1, c_2, \ldots, c_m)$, where c_i are target qualifying constraints
- $A = (A_1 = (a_1, a_2), A_2 = (a'_1, a'_2))$, where a_i and a'_i are variables

A query is constructed for each target qualifying constraint (i.e., m queries) where the SELECT clause is formed by using variables in A_1 (e.g., SELECT $?a_1?a_2$). Within each query, the constraint $c_i \in C$ can be mapped to a FILTER clause or to a constraint in a WHERE clause of a SPARQL query. In order to map the constraints to a WHERE clause, a target qualifying triple pattern is constructed for each constraint. Assuming vpt_1 contains data to which a constraint can be applied, the target qualifying triple pattern is specified as ($?var :vpt_1 c_i$),

```
SELECT ?movieName ?rating          SELECT ?movieName ?rating
WHERE { ?user hasName: "user8".    WHERE { ?user hasName: "user34".
        ?user hasRated: ?movie.            ?user hasRated: ?movie.
        ?movie hasMName: ?movieName.       ?movie hasMName: ?movieName.
        ?movie hasRating: ?rating. }       ?movie hasRating: ?rating. }
```

(a) TQ Queries

```
SELECT ?movieName ?date
WHERE   {   ?movie hasName: ?movieName.
            ?movie hasDate: ?date. }
```

(b) TD Query

Fig. 11. Queries used to construct TD and TQ tables

where $?var$ is some variable. Moreover, the remaining tables $vpt_2 \ldots vpt_m$ are joined together and with the intermediate result of the target qualifying triple pattern. A similar method can be applied if a FILTER clause is preferred. Instead of providing a constraint in the target qualifying triple pattern, a new variable is introduced, as in $(?var : vpt_1\ ?constraint)$. This constraint variable is then used in the FILTER clause along with the actual constraint to filter out unwanted results. An example FILTER clause is FILTER regex(?constraint, c_1).

To illustrate this process, consider Query 1 and the ontology depicted in Fig. 18. Suppose the person issuing the query is interested in the following movie-raters: $user8$ and $user34$. Since the query is requesting movies (i.e., the name of the movies) whose rating is maximized, we define $A = (?movieName, ?rating)$ because rating depends on the movie and the movie-rater. In addition, by examining Fig. 18, we have the following VPTs: $hasName, hasRated, hasMName, hasRating$. Since vpt_1 is $hasName$, we apply each constraint to this table by using a target qualifying triple pattern, such as $(?user\ hasName$ "user8"). Therefore, given C and A_1, the two queries in Fig. 12a are constructed, which produces the TQ tables.

A similiar approach is used to form the TD table. Since no target qualifying constraints are needed in this case, no FILTER condition is required and the only argument of interest is A_2, which contains variables that will be listed in the SELECT clause. The WHERE claused is formed by joining $vpt_{m+1}, \ldots vpt_n$. Continuing from the previous example, we have the following VPTs: $hasName, hasDate$, and $A_2 = (?movieName, ?date)$. Therefore, the query in Fig. 12b is constructed.

In order to retrieve data from Sesame's repository, we implemented a server-side $SPARQL\ adapter$ whose primary task is issuing SPARQL queries and serves as an intermediary between Sesame and $SkyPackage$. The adapter utilizes Sesame's $HTTPRepsoitory$ component to execute the three queries and to retrieve the results. The results of each query are then stored in a data structure for processing and sent to $SkyPackage$ along with the list of preferences PF. After $SkyPackage$ is performed on the data returned from the adapter, the results are presented to the user. Figure 11 provides a high-level overview indicating the main components of our framework.

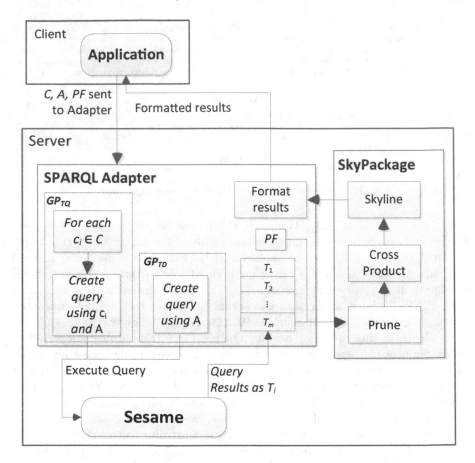

Fig. 12. Framework with SPARQL adapter

6   Evaluation

The main goal of this evaluation was to compare the performance of the proposed algorithms using both synthetic and real datasets with respect to package size scalability. In addition we compared the the feasibility of answering the skyline package problem using the VPT storage model and the TDTQ storage model.

6.1   Setup

All experiments were conducted on a Linux machine with a 2.33GHz Intel Xeon processor and 40GB memory, and all algorithms were implemented in Java SE 6. All data used was converted to RDF format using the Jena API and stored in Oracle Berkeley DB.

We compared four algorithms, *JCPS*, *SkyJCPS*, *RSJFH-CPS*, and *SkyPackage*. During the skyline phase of each of these algorithms, we used the *block-nested-loops* (*BNL*) [10] algorithm. The package size metric was used for the scalability study of the algorithms. Since the Cartesian product phase is likely to be the dominant cost in skyline package queries, it is important to analyze how the algorithms perform when the package size grows, which increases the input size of the Cartesian product phase.

6.2   Synthetic Data

Dataset. Since we are unaware of any RDF data generators that allow generation of different data distributions, the data used in the evaluations were generated using a synthetic data generator [2]. The data generator produces relational data in different distributions, which was converted to RDF using the Jena API. We generated three types of data distributions: correlated, anti-correlated, and normal distributions. For each type of data distribution, we generated datasets of different sizes and dimensions.

Data Size Scalability. The first evaluation was performed to compare execution time among the three algorithms within the same package size using the three data distributions. The data consisted of triple sizes ranging from 450 to 635 and package sizes ranging from 2 to 5. While this may appear to be orders of magnitudes smaller than traditional evaluation corpora, it is important to note that the search space for package queries grows more aggressively than that of traditional pattern matching. We chose this triple size range to ensure that packages of different sizes can easily be compared and also to ensure that evaluation results would come in a reasonable time for larger package sizes. An increase in package size implies an increase in the number of tables, which also implies more Cartesian products. To ensure that the triple size remained approximately the same across different package sizes, we reduced the number of tuples in each table as the package size increased. Figure 13 shows the triple size and the number of tuples in each table for each package size as well as the approximate Cartesian product size.

While a triple size of 635 may seem small, Fig. 13 indicates that this triple size yields approximately 52.5 M tuples for a package size of 5 after a Cartesian product is performed. We were unable to obtain any results for triple size 635 using a package size of 5 for *JCPS*, as it ran for hours on this dataset. Figures 14 and 15 show the results and are plotted using a logarithmic scale. No anomalies were found within packages of the same size.

Package Size Scalability. In Fig. 16, we show how the algorithms perform across packages of size 2 to 5 for the a triple size of approximately 635 triples. Due to the exponential increase of the Cartesian product phase, this triple size is the largest possible in order to evaluate all three algorithms. For all package sizes, *SkyJCPS* performs better than *JCPS* because of the initial skyline algorithm performed to reduce the input size of the Cartesian product phase. *RSJFH* outperformed *SkyJCPS* for packages of size 5. We argue that *SkyJCPS* may perform slightly slower than *RSJFH* on small datasets distributed among many tables. In this scenario, *SkyJCPS* has six tables to examine, while *RSJFH* has

Package Size	Triple Size	Tuples in each table	Approx Cartesian Product Size	Package Size	Triple Size	Tuples in each table	Approx Cartesian Product Size
5	635	35	52.5M	3	639	53	149,000
	599	33	39M		603	50	125,000
	545	30	24M		543	45	91,000
	509	28	17M		507	42	74,000
	455	25	9.7M		447	37	50,000
4	634	42	3.1M	2	632	70	4,900
	604	40	2.5M		605	67	4,500
	544	36	1.7M		542	60	3,600
	499	33	1.2M		497	55	3,000
	454	30	810,000		452	50	2,500

Fig. 13. Synthetic data triple sizes

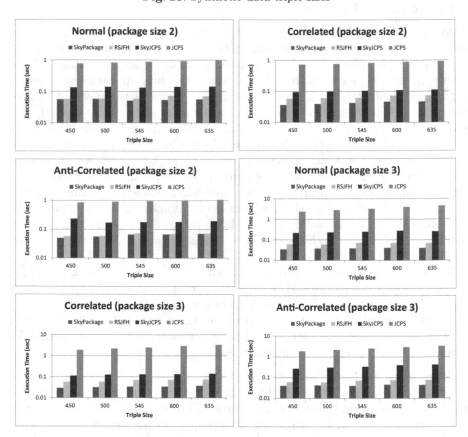

Fig. 14. Evaluation results for package sizes 2 and 3

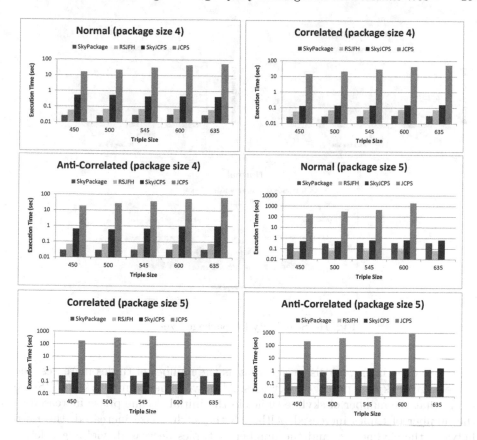

Fig. 15. Evaluation results for package sizes 4 and 5

only two tables. Evaluation results from the real datasets, which is discussed next, ensure us that *SkyJCPS* significantly outperforms *RSJFH* when the dataset is large and distributed among many tables. Due to the logarithmic scale used, it may seem that some of the algorithms have the same execution time for equal sized packages. This is not the case, and since *BNL* was the single-table skyline algorithm used, the algorithms performed best using correlated data and worst using anti-correlated data.

Average Prunability. To evaluate *SkyPackage*'s prunability, we collected the number of tuples that entered the Cartesian product phase and compared it to the total number of initial tuples for each data distribution and triple size. We then took the average over the three data distributions. The average prunability results can be seen in Fig. 17. As the package size increases, a larger percentage of tuples enters the Cartesian product phase. Even though the triple size remains approximately the same in all packages, the total number of tuples increases as the package size grows. For example, a triple size of 635 for a package of size 2 consists of 140 tuples, whereas a package of size 5 has 175 tuples (a 25 % increase).

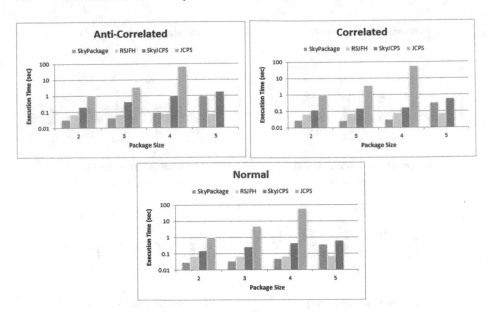

Fig. 16. Scalability for package sizes 2–5

From our experiment, we discovered that as the number of tuples increases, the percentage of skyline packages decreases. Also, while performing $SkyJCPS$, we found that the number of skyline tuples in the initial skyline phase increased as the number of tuples increased. Although the skyline size increased, the ratio between the skyline size and the number of tuples decreased, yielding a lower percentage. We argue that this is also the case with $SkyPackage$.

6.3 MovieLens Dataset

The first real-world dataset evaluated was MovieLens[2], which consisted of 10 million ratings and 10,000 movies.

Package-size Scalability and Prunability. We randomly chose a subset of users, with partiality to those who have rated a large number of movies, from the dataset for use in our package-size evaluations. The users consisted of those with IDs 8, 34, 36, 65, and 215, who rated 800, 639, 479, 569, and 1,242 movies, respectively. We used Query 1, where $n = 3, 4, 5$ for evaluation. The packages of size 3 consisted of users with IDs 8, 34, and 36, packages of size 4 included those three as well as the user with ID 65, and packages of size 5 included all five users. In formal notation, the query where $n = 3$ (a similar query is used for $n = 4, 5$) is $GP_{[r,X,\mathcal{F},PF]}$, where

---

[2] http://www.grouplens.org/node/73/

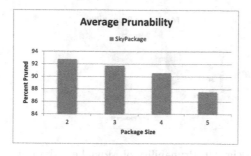

Fig. 17. Prunability of synthetic data

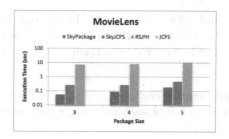

Fig. 18. Package size scalability for MovieLens

- $r = ?movie$
- $X = \{(?movieRater, hasName, \text{"8"}), (?movieRater, hasName, \text{"34"}),$
  $(?movieRater, hasName, \text{"36"}) \}$
- $\mathcal{F} = \{SR = average()_{rating}\}$
- $PF = \{(SR; \prec_{max})\}$

The results of this experiment can be seen in Fig. 18. It is easily observed that *SkyPackage* performed better in all cases. We were unable to obtain any results from *JCPS* as it ran for hours. The next worst performing algorithm was *RSJFH*, followed by *SkyJCPS*. Due to the number of joins required to construct the tables in the format required by *RSJFH*, most of its time was spent during the initial phase, i.e., before the Cartesian product phase. Figure 19 shows the percent of tuples pruned by *SkyPackage*. In all three package sizes, approximately 99 % of the tuples were pruned.

6.4   Book-Crossing Dataset

The next real dataset used for evaluations was Book-Crossing[3], which consists of approximately 271,000 books rated by approximately 278,000 users.

Package-size Scalability and Prunability. Using a similar approach as we did with the MovieLens dataset, we randomly chose a subset of users for evaluating packages of different sizes. The users consisted of those with IDs 11601,

---

[3] http://www.informatik.uni-freiburg.de/~cziegler/BX/dataset

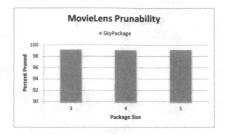

Fig. 19. Prunability of MovieLens dataset

11676, 16795, 23768, and 23902, who rated 1,571, 13,602, 2,948, 1,708, and 1,503 books, respectively. The target descriptive table contained approximately 271,000 tuples, i.e., all the books. We used Query 2, where $n = 3, 4, 5$ for evaluation.

Query 2. *Given n book-raters, find packages of n books such that the average rating of all the books is high and each book-rater has rated at least one of the books.*

In formal notation, the query where $n = 3$ (a similar query is used for $n = 4, 5$) is $GP_{[r,X,\mathcal{F},PF]}$, where

- $r = ?book$
- $X = \{(?bookRater, hasName, \text{"11601"}), (?bookRater, hasName, \text{"11676"}), (?bookRater, hasName, \text{"16795"}) \}$
- $\mathcal{F} = \{SR = average()_{rating}\}$
- $PF = \{(SR; \prec_{max})\}$

The packages of size 3 consisted of users with IDs 11601, 11676, and 16795, packages of size 4 included those three as well as the user with ID 23768, and packages of size 5 included all five users. The results of this experiment can be seen in Fig. 20. The Book-Crossing dataset followed the same performance pattern as the MovieLens datasets, i.e., *SkyPackage* performed the best while *RSJFH* performed the worst. Although the overall execution time of the Book-Crossing dataset was longer than the MovieLens dataset, the data we used from the Book-Crossing dataset consisted of more tuples. Fig. 21 shows the percent of tuples pruned by *SkyPackage* from the Book-Crossing dataset.

6.5   Storage Model Evaluation

For each of the above experiments, we evaluated our storage model by comparing the time it took to load the RDF file into the database using our storage model versus using VPTs. All data was indexed using a B-trees.

For the synthetic data, we took the average time over the three data distributions of the same package size. Figure 22a shows the results for packages of

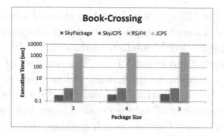

Fig. 20. Package size scalability for Book-Crossing

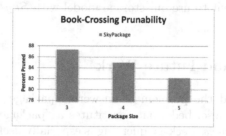

Fig. 21. Prunability of Book-Crossing dataset

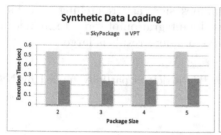

(a) Synthetic Data

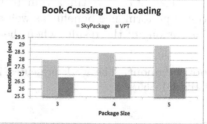

(b) Book-Crossing Data

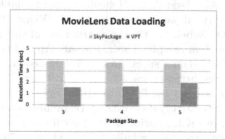

(c) MovieLens Data

Fig. 22. Database build

size 2, 3, 4, and 5. In Fig. 22b, c, we show the time of inserting the MovieLens and the Book-Crossing datasets, respectively, for packages of size 3, 4, and 5.

For both the synthetic and MovieLens datasets, loading the data using our storage model took longer than using VPTs. The number of tables created using both approaches are not always equal, and either approach could have more tables than the other. Since the time to load the data is roughly the same for each package size, the number of tables created does not necessarily have that much effect on the total time. Our approach imposes additional time because of the triple patterns that must be matched, as explained in Sect. 4.1. Since the time difference between the two is small and the database only has to be built once, it is more efficient to use our storage model with *SkyPackage* than using VPTs.

7   Conclusion and Future Work

This article addressed the problem of answering package skyline queries. We have formalized and described what constitutes a "package" and have defined the term *skyline packages*. Package querying is especially useful for cases where a user requires multiple objects to satisfy certain constraints. We introduced three algorithms for solving the package skyline problem. Future work will consider the use of additional optimization techniques such as prefetching to achieve additional performance benefits as well as the integration of top-k techniques to provide ranking of the results when the size of query result is large.

Acknowledgment. The work presented in this article is partially funded by NSF grant IIS-0915865.

References

1. Abadi, D., Marcus, A., Madded, S., Hollenbach, K.: Scalable semantic web data management using vertical partitioning. In: VLDB, Vienna (2007)
2. Barndorff-Nielsen, O., Sobel, M.: On the distribution of the number of admissable points in a vector random sample. Theor. Probab. Appl. **11**(2), 249–269 (1996)
3. Beckett, D.: RDFXML syntax specification. Recommendation, World Wide Web Consortium (2004). See http://www.w3.org/TR/rdf-syntax-grammar/
4. Beckett, D., Berners-Lee, T.: Turtle - Terse RDF triple language. World Wide Web Consortium (2011). See http://www.w3.org/TeamSubmission/turtle/
5. Bentley, J.L., Clarkson, K.L., Levine, D.B.: Fast linear expected-time algorithms for computing maxima and convex hulls. In: SODA, pp. 179–187 (1990)
6. Bentley, J.L., Kung, H.T., Schkolnick, M., Thompson, C.D.: On the average number of maxima in a set of vectors and applications, pp. 536–543. ACM (1978)
7. Berners-Lee, T., Fielding, R., Irvine, U.C., Masinter, L.: Uniform resource identifiers. (URI), Generic Syntax. IETF (1998)
8. Berners-Lee, T., Hendler, J., Lassila, O.: The semantic web. Scientif. Am. **284**(5), 34–43 (2001)

9. Boag, S., Chamberlin, D., Fernandez, M.F., Florescu, D., Robie, J., Simeon, J.: XQuery 1.0: An XML Query Language, 2nd edn. Recommendation, World Wide Web Consortium (2010). See http://www.w3.org/TR/xquery/

10. Borzsonyi, S., Kossmann, D., Stocker, K.: The Skyline operator. In: ICDE, Heidelberg, pp. 421–430 (2001)

11. Bray, T., Paoli, J., Sperberg-McQueen, C.M., Maler, E., Yergeau, F.: Extensible Markup Language (XML) 1.0. Recommendation, World Wide Web Consortium (2000). See http://www.w3.org/TR/2008/REC-xml-20081126/

12. Brickley, D., Guha, R.V.: Resource Description Framework (RDF) Schema Specification 1.0 Canditate Recommendation, World Wide Web Consortium (2000). See http://www.w3.org/TR/2000/CR-rdf-schema-20000327

13. Broekstra, J.: SeRQL: Sesame RDF query language. SWAD-Europe, pp. 55–68 (2003)

14. Broekstra, J., Kampman, A., Harmelen, F.V.: Sesame: A generic architecture for storing and querying RDF and RDF schema. In: ISWC, pp. 54–68 (2012)

15. Carrol, J., McBride, B.: The Jena semantic web toolkit. Public API, HP-Labs, Bristol (2001). See http://www.hpl.hp.com/semweb/jena-top.html

16. Chen, L., Gao, S., Anyanwu, K.: Efficiently evaluating skyline queries on RDF databases. In: Antoniou, G., Grobelnik, M., Simperl, E., Parsia, B., Plexousakis, D., De Leenheer, P., Pan, J. (eds.) ESWC 2011, Part II. LNCS, vol. 6644, pp. 123–138. Springer, Heidelberg (2011)

17. Chomicki, J.: Preference formulas in relational queries. TODS 28(4), 427–466 (2003)

18. Chomicki, J., Godfrey, P., Gryz, J., Liang, D.: Skyline with presorting. In: ICDE, pp. 717–719 (2003)

19 Grant, J., Beckett, D.: RDF test cases. Recommendation, World Wide Web Consortium (2004). See http://www.w3.org/TR/rdf-testcases/#ntriples

20. Deng, T., Fan, W., Geerts, F.: On the complexity of package recommendation problems. PODS, pp. 261–272 (2012)

21. Feigenbaum, L., Williams, G.T., Clark, K.G., Torress Hayes, E.: SPARQL 1.1 Protocol. Working draft. World Wide Web Consortium (2001). See http://www.w3.org/TR/sparql11-protocol/

22. Karvounarakis, G., Alexaki, S., Christophides, V., Plexousakis, D., Scholl, M.: RQL: A declarative query language for RDF. World Wide Web, pp. 592–603 (2002)

23. Khabbaz, M., Lakshmanan, L.V.S.: TopRecs: Top-k algorithms for item-based collaborative filtering. EDBT, 213–224 (2011)

24. Khalefa, M.E., Mokbel, M.F., Levandoski, J.J.: Skyline query processing for incomplete data (2008)

25. KieBling, W.: Foundations of preferences in database systems. In: VLDB, pp. 311–322 (2002)

26. Kossmann, D., Ramsak, F., Rost, S.: Shooting stars in the sky: An online algorithm for skyline queries. In: VLDB, pp. 275–286 (2002)

27. Koudas, N., Li, C., Tung, A.K.H., Vernica, R.: Relaxing join and selection queries. In: VLDB (2006)

28. Kung, H.T., Luccio, F., Preparata, F.P.: On finding the maxima of a set of vectors. JACM 22(4), 469–476 (1975)

29. Lassila, O., Swick, R.R.: Resource description framework (RDF): Model and syntax specification. Recommendation, World Wide Web Consortium (1999). See http://www.w3.org/TR/REC-rdf-syntax/

30. Berners-Lee, T., Connolly, D.: Notation3 (N3): A readable RDF syntax. World Wide Web Consortium (2011). See http://www.w3.org/TeamSubmission/n3/

31. McLain, D.H.: Drawing contours from arbitrary data points. Comput. J. **17**(4), 318–324 (1974)
32. Raghavan, V., Rundensteiner, E.: SkyDB: Skyline aware query evaluation framework. In: IDAR (2009)
33. Papadias, D., Tao, Y., Fu, G., Seeger, B.: Progressive skyline computation in database systems. TODS **24**(2), 41–82 (2005)
34. Preparata, F., Shamos, M.: Computational Geometry: An Introduction. Springer, Heidelberg (1985)
35. Seaborne, A.: RDQL - A query language for RDF. World Wide Web Consortium (2004). See http://www.w3.org/Submission/RDQL/
36. Sessoms, M., Anyanwu, K.: SkyPackage: From finding items to finding a skyline of packages on the semantic web. Proceedings of the JIST (to appear), (2012)
37. Shah, K., Gadge, J.: Semantic web services for E learning: Engineering and technology domain. In: IJCTE 2001, pp. 727–731 (2001)
38. Siberski, W., Pan, J.Z., Thaden, U.: Querying the semantic web with preferences. In: Cruz, I., Decker, S., Allemang, D., Preist, C., Schwabe, D., Mika, P., Uschold, M., Aroyo, L.M. (eds.) ISWC 2006. LNCS, vol. 4273, pp. 612–624. Springer, Heidelberg (2006)
39. Sintek, M., Decker, S.: TRIPLE–A query, inference, and transformation language for the semantic web. In: Horrocks, I., Hendler, J. (eds.) ISWC 2002. LNCS, vol. 2342, p. 364. Springer, Heidelberg (2002)
40. Souzis, A.: RxPath specification proposal. See http://rx4rdf.liminalzone.org/RxPathSpec
41. Tan, K.L., Eng, P.K., Ooi, B.C.: Efficient progressive skyline computation. In: VLDB, pp. 301–310 (2001)
42. Uschold, M., Gruninger, M.: Ontologies: principles, methods and applications. Knowl. Eng. Rev. **11**(2), 93–136 (1996)
43. Vlachou, A., Doulkeridis, C., Polyzotis,N.: Skyline query processing over joins. In: SIGMOD, pp. 73–84 (2011)
44. Xie, M. Lakshmanan, L.V.S., Wood, P.T.: Breaking out of the box of recommendations: from items to packages. In: RecSys, pp. 151–158 (2010)
45. Xie, M. Lakshmanan, L.V.S., Wood, P.T.: CompRec-Trip: a composite recommendation system for travel planning. In: ICDE, pp. 1352–1355 (2011)
46. Yiu, M.L., Mamoulis, N.: Efficient processing of top-k dominating queries on multi-dimensional data. In: VLDB, pp. 483–494 (2007)
47. Jin, W., Ester, M., Hu, Z., Han, J.: The Multi-relational skyline operator. In: ICDE, pp. 1276–1280 (2007)
48. Guo, X., Xiao, C., Ishikawa, Y.: Combination skyline queries. In: Hameurlain, A., Küng, J., Wagner, R., Liddle, S.W., Schewe, K.-D., Zhou, X. (eds.) Transactions on Large-Scale Data- and Knowledge-Centered Systems VI. LNCS, vol. 7600, pp. 1–30. Springer, Heidelberg (2012)

SemLAV: Local-As-View Mediation
for SPARQL Queries

Gabriela Montoya[1]([✉]), Luis-Daniel Ibáñez[1], Hala Skaf-Molli[1], Pascal Molli[1], and Maria-Esther Vidal[2]

[1] LINA– Nantes University, Nantes, France
{gabriela.montoya,luis.ibanez,hala.skaf,pascal.molli}@univ-nantes.fr
[2] Universidad Simón Bolívar, Caracas, Venezuela
mvidal@ldc.usb.ve

Abstract. The Local-As-View (LAV) integration approach aims at querying heterogeneous data in dynamic environments. In LAV, data sources are described as views over a global schema which is used to pose queries. Query processing requires to generate and execute query rewritings, but for SPARQL queries, the LAV query rewritings may not be generated or executed in a reasonable time.

In this paper, we present SemLAV, an alternative technique to process SPARQL queries over a LAV integration system without generating rewritings. SemLAV executes the query against a partial instance of the global schema which is built on-the-fly with data from the relevant views. The paper presents an experimental study for SemLAV, and compares its performance with traditional LAV-based query processing techniques. The results suggest that SemLAV scales up to SPARQL queries even over a large number of views, while it significantly outperforms traditional solutions.

Keywords: Semantic Web · Data integration · Local-as-view · SPARQL query

1 Introduction

Processing queries over a set of autonomous and semantically heterogeneous data sources is a challenging problem. Particularly, a great effort has been done by the Semantic Web community to integrate datasets into the Linked Open Data (LOD) cloud [1] and make these data accessible through SPARQL endpoints which can be queried by federated query engines. However, there are still a large number of data sources and Web APIs that are not part of the LOD cloud. As consequence, existing federated query engines cannot be used to integrate these data sources and Web APIs. Supporting SPARQL query processing over these environments would extend federated query engines into the deep Web.

Gabriela Montoya—Unit UMR6241 of the Centre National de la Recherche Scientifique (CNRS).

A. Hameurlain et al. (Eds.): TLDKS XIII, LNCS 8420, pp. 33–58, 2014.
DOI: 10.1007/978-3-642-54426-2_2, © Springer-Verlag Berlin Heidelberg 2014

Two main approaches exist for data integration: data warehousing and mediators. In data warehousing, data are transformed and loaded into a repository; this approach may suffer from freshness problem [2]. In the mediator approach, there is a global schema over which the queries are posed and views describe data sources. Three main paradigms are proposed: Global-As-View (GAV), Local-As-View (LAV) and Global-Local-As-View (GLAV). In GAV mediators, entities of the global schema are described using views over the data sources, including or updating data sources may require the modification of a large number of views [3]. Whereas, in LAV mediators, the sources are described as views over the global schema, adding new data sources can be easily done [3]. Finally, GLAV is a hybrid approach that combines both LAV and GAV approaches. GAV is appropriate for query processing in stable environments. A LAV mediator relies on a query rewriter to translate a mediator query into the union of queries against the views. Therefore, it is more suitable for environments where data sources frequently change. Despite of its expressiveness and flexibility, LAV suffers from well-known drawbacks: (i) existing LAV query rewriters only manage conjunctive queries, (ii) the query rewriting problem is NP-complete for conjunctive queries, and (iii) the number of rewritings may be exponential.

SPARQL queries exacerbate LAV limitations, even in presence of conjunctions of triple patterns. For example, in a traditional database system, a LAV mediator with 140 conjunctive views can generate 10,000 rewritings for a conjunctive query with eight subgoals [4]. In contrast, the number of rewritings for a SPARQL query can be much larger. SPARQL queries are commonly comprised of a large number of triple patterns and some may be bound to *general predicates* of the RDFS or OWL vocabularies, e.g., rdf:type, owl:sameAs or rdfs:label, which are usually used in the majority of the data sources. Additionally, queries can be comprised of several star-shaped sub-queries [5]. Finally, a large number of variables can be projected out. All these properties impact the complexity of the query rewriting problem, even enumerating query rewritings can be unfeasible. For example, a SPARQL query with 12 triple patterns that comprises three star-shaped sub-queries can be rewritten using 476 views in billions of rewritings. This problem is even more challenging considering that statistics may be unavailable, and there are no clear criteria to rank or prune the generated rewritings [6]. It is important to note that for conjunctive queries, GLAV query processing tasks are at least as complex as LAV tasks [7].

In this paper, we focus on the LAV approach, and propose SemLAV, the first scalable LAV-based approach for SPARQL query processing. Given a SPARQL query Q on a set M of LAV views, SemLAV selects relevant views for Q and ranks them in order to maximize query results. Next, data collected from selected views are included into a partial instance of the global schema, where Q can be executed whenever new data is included; and thus, SemLAV incrementally produces query answers. Compared to a traditional LAV approach, SemLAV avoids generating rewritings which is the main cause of the combinatorial explosion in traditional rewriting-based approaches; SemLAV also supports the execution of SPARQL queries. The performance of SemLAV is no more dependent on the number of rewritings, but it does depend on the number and size of relevant views. Space required to temporarily include relevant views in the global schema instance may be considerably larger than the space required to execute all the query rewritings one by one. Nevertheless, executing the query once on the partial instance of the global schema could produce the answers obtained by executing all the rewritings.

To empirically evaluate the properties of SemLAV, we conducted an experimental study using the Berlin Benchmark [8] and queries and views designed by Castillo-Espinola [9]. Results suggest that SemLAV outperforms traditional LAV-based approaches with respect to answers produced per time unit, and provides a scalable LAV-based solution to the problem of executing SPARQL queries over heterogeneous and autonomous data sources.

The contributions of this paper are the following:

- Formalization of the problem of finding the set of relevant LAV views that maximize query results; we call this problem MaxCov.
- A solution to the MaxCov problem.
- A scalable and effective LAV-based query processing engine to execute SPARQL queries, and to produce answers incrementally.

The paper is organized as follows: Sect. 2 presents basic concepts, definitions and a motivating example. Section 3 defines the MaxCov problem, SemLAV query execution approach and algorithms. Section 4 reports our experimental study. Section 5 summarizes related work. Finally, conclusions and future work are outlined in Sect. 6.

2   Preliminaries

Mediators are components of the mediator-wrapper architecture [10]. They provide an uniform interface to autonomous and heterogeneous data sources. Mediators also rewrite an input query into queries against the data sources, and merge data collected from the selected sources. Wrappers are software components that solve interoperability between sources and mediators by translating data collected from the sources into the schema and format understood by the mediators; the schema exposed by the wrappers is part of the schema exposed by its corresponding mediator.

The problem of processing a query Q over a set of heterogeneous data sources corresponds to answer Q using the instances of these sources. Although this problem has been extensively studied by the Database community [11], it has not been addressed for SPARQL queries. The following definitions are taken from Database existing solutions. Many of them are given for conjunctive queries. A conjunctive query has the form: $Q(\bar{X})$:- $p_1(\bar{X}_1), \ldots, p_n(\bar{X}_n)$, where p_i is a predicate, $\bar{X}_i$ is a list of variables and constants, $Q(\bar{X})$ is the head of the query, $p_1(\bar{X}_1), \ldots, p_n(\bar{X}_n)$ is the body of the query, and each element of the body is a query subgoal. In a conjunctive query, distinguished variables are variables that appear in the head, they should also appear in the body. Variables that appear in the body, but not in the head are existential variables.

Definition 1 (LAV Integration System [12]). *A LAV integration system is a triple IS=< G, S, M > where G is a global schema, S is a set of sources or source schema, and M is a set of views that map sources in S into the global schema G.*

For the rest of the paper, we assume that views in *M* are limited to conjunctive queries. Both views and mediator queries are defined over predicates in *G*.

Definition 2 (Sound LAV View [12]). *Given IS=< G, S, M > a LAV integration system, and a view v_i in M. The view v_i is sound if for all instance $I(v_i)$ of v_i, and all D virtual database instance of G, $I(v_i)$ is contained in the evaluation of view v_i over D, i.e., $I(v_i) \subseteq v_i(D)$.*

Definition 3 (Query Containment and Equivalence [13]). *Given two queries Q1 and Q2 with the same number of arguments in their heads, Q1 is contained in Q2, $Q1 \sqsubseteq Q2$, if for any database instance D the answer of Q1 over D is contained in the answer to Q2 over D, $Q1(D) \subseteq Q2(D)$. Q1 is equivalent to Q2 if $Q1 \sqsubseteq Q2$ and $Q2 \sqsubseteq Q1$.*

Definition 4 (Containment Mapping [13]). *Given two queries Q1 and Q2, $\bar{X}$ and $\bar{Y}$ the head variables of Q1 and Q2 respectively, and ψ a variable mapping from Q1 to Q2, ψ is a containment mapping if $\psi(\bar{X}) = \bar{Y}$ and for every query subgoal $g(\bar{X}_i)$ in the body of Q1, $\psi(g(\bar{X}_i))$ is a subgoal of Q2.*

Theorem 1 (Containment [13]). *Let Q1 and Q2 be two conjunctive queries, then there is a containment mapping from Q1 to Q2 if and only if $Q2 \sqsubseteq Q1$.*

Definition 5 (Query Unfolding [13]). *Given a query Q and a query subgoal $g_i(\bar{X}_i)$, $g_i(\bar{X}_i) \in body(Q)$, where g_i corresponds to a view: $g_i(\bar{Y})$:-$s_1(\bar{Y}_1), \dots, s_n(\bar{Y}_n)$, the unfolding of g_i in Q is done using a mapping τ from variables in $\bar{Y}$ to variables in $\bar{X}_i$, replacing $g_i(\bar{X}_i)$ by $s_1(\tau(\bar{Y}_1)), \dots, s_n(\tau(\bar{Y}_n))$ in Q. Variables that occur in the body of g_i but not in $\bar{X}_i$ are replaced by fresh (unused) variables by mapping τ.*

Definition 6 (Equivalent Rewriting [11]). *Let Q be a query and $M = \{v_1, \dots, v_m\}$ be a set of views definitions. The query Q' is an equivalent rewriting of Q using M if:*

– *Q' refers only to views in M, and*
– *Q' is equivalent to Q.*

Definition 7 (Maximally-Contained Rewriting [11]). *Let Q be a query, $M = \{v_1, \dots, v_m\}$ be a set of views definitions, and L be a query language[1]. The query Q' is a maximally-contained rewriting of Q using M with respect to L if:*

– *Q' is a query in L that refers only to the views in M,*

---

[1] *L is a query language defined over the alphabet composed of the global and source schema.*

- Q' is contained in Q, and
- there is no rewriting $Q_1 \in L$, such that $Q' \sqsubseteq Q_1 \sqsubseteq Q$ and Q_1 is not equivalent to Q'.

Theorem 2 (Number of Candidate Rewritings [2]). *Let N, O and M be the number of query subgoals, the maximal number of views subgoals, and the set of views, respectively. The number of candidate rewritings in the worst case is:* $(O \times |M|)^N$.

Theorem 3 (Complexity of Finding Rewritings [11]). *The problem of finding an equivalent rewriting is NP-complete.*

Consider L in Definition 7 as the union of conjunctive queries, then view v would be used to answer query Q if there is one conjunctive query $r \in Q'$ such that v appears as the relation of one of r query subgoals. As $Q' \sqsubseteq Q$, then $r \sqsubseteq Q$. View v is called a relevant view atom. The next definition formalizes this notion.

Definition 8 (Relevant View Atom [13]). *A view atom v is relevant for a query atom g if one of its subgoals can play the role of g in the rewriting. To do that, several conditions must be satisfied: (1) the view subgoal should be over the same predicate as g, and (2) if g includes a distinguished variable of the query, then the corresponding variable in v must be a distinguished variable in the view definition.*

The concepts of relevant view and coverage have been widely used in the literature [11,13]; nevertheless, they have been introduced in an informal way. The following definitions precise the properties that are assumed in this paper.

Definition 9 (Relevant Views). *Let Q be a conjunctive query, $M = \{v_1, \ldots, v_m\}$ be a set of view definitions, and q be a query subgoal, i.e., $q \in body(Q)$. The set of relevant views for q corresponds to the set of relevant view atoms for the query subgoal q, i.e., $RV(M,q) = \{\tau(v) : v \in M \wedge w \in body(v) \wedge \psi(q) = \tau(w) \wedge (\forall x : x \in Vars(q) \wedge distinguished(x,Q) : distinguished(x,v))\}^2$. The set of relevant views for Q corresponds to the views that are relevant for at least one query subgoal, i.e., $RV(M,Q) = \{\tau(v) : q \in body(Q) \wedge v \in M \wedge w \in body(v) \wedge \psi(q) = \tau(w) \wedge (\forall x : x \in Vars(q) \wedge distinguished(x,Q) : distinguished(x,v))\}$.*

Definition 10 (Coverage). *Let Q be a conjunctive query, v be a view definition, q be a query subgoal, and w be a view subgoal. The predicate $covers(w,q)$ holds if and only if w can play the role of q in a query rewriting.*

We illustrate some of the given definitions for the LAV-based query rewriting approach using SPARQL queries. This will provide evidence of the approach limitations even for simple queries. In the following example, the global schema G is defined over the Berlin Benchmark [8] vocabulary. Consider a SPARQL query Q on G; Q has seven subgoals and returns information about products as

---

[2] $\psi(q)$ corresponds to the application of ψ to the variables of q (idem for $\tau(w)$).

shown in Listing 1.1. Listing 1.3 presents Q as a conjunctive query, where triple patterns are represented as query subgoals.

```
SELECT *
WHERE {
  ?X1 rdfs:label ?X2 .
  ?X1 rdfs:comment ?X3 .
  ?X1 bsbm:productPropertyTextual1 ?X8 .
  ?X1 bsbm:productPropertyTextual2 ?X9 .
  ?X1 bsbm:productPropertyTextual3 ?X10 .
  ?X1 bsbm:productPropertyNumeric1 ?X11 .
  ?X1 bsbm:productPropertyNumeric2 ?X12 .
}
```

Listing 1.1. SPARQL query Q

```
SELECT *
WHERE {
  ?X1 rdfs{:}label ?X2 .
  ?X1 rdf{:}type ?X3 .
  ?X1 bsbm{:}productFeature ?X4
}
```

Listing 1.2. SPARQL View $s1$

```
Q(X1, X2, X3, X8, X9, X10, X11, X12) :- label(X1, X2), comment(X1, X3),
      productPropertyTextual1(X1, X8), productPropertyTextual2(X1, X9),
      productPropertyTextual3(X1, X10), productPropertyNumeric1(X1, X11),
      productPropertyNumeric2(X1, X12)
```

Listing 1.3. Q expressed as a conjunctive query

```
s1(X1,X2,X3,X4):-label(X1,X2),type(X1,X3),productfeature(X1,X4)
s2(X1,X2,X3):-type(X1,X2),productfeature(X1,X3)
s3(X1,X2,X3,X4):-producer(X1,X2),label(X2,X3),publisher(X1,X2),
      productfeature(X1,X4)
s4(X1,X2,X3):-productfeature(X1,X2),label(X2,X3)
s5(X1,X2,X3,X4,X5,X6,X7):-label(X1,X2),comment(X1,X3),producer(X1,X4),
      label(X4,X5),publisher(X1,X4),productpropertytextual1(X1,X6),
      productpropertynumeric1(X1,X7)
s6(X1,X2,X3,X4,X5):-label(X1,X2),product(X3,X1),price(X3,X4),vendor(X3,X5)
s7(X1,X2,X3,X4,X5,X6):-label(X1,X2),reviewfor(X3,X1),reviewer(X3,X4),
      name(X4,X5),title(X3,X6)
s9(X1,X2,X3,X4):-reviewfor(X1,X2),title(X1,X3),text(X1,X4)
s10(X1,X2,X3):-reviewfor(X1,X2),rating1(X1,X3)
s11(X1,X2,X3,X4,X5,X6,X7):-label(X1,X2),comment(X1,X3),producer(X1,X4),
      label(X4,X5),publisher(X1,X4),productpropertytextual2(X1,X6),
      productpropertynumeric2(X1,X7)
s12(X1,X2,X3,X4,X5,X6,X7):-label(X1,X2),comment(X1,X3),producer(X1,X4),
      label(X4,X5),publisher(X1,X4),productpropertytextual3(X1,X6),
      productpropertynumeric3(X1,X7)
s13(X1,X2,X3,X4,X5,X6,X7):-label(X1,X2),product(X3,X1),price(X3,X4),
      vendor(X3,X5),offerwebpage(X3,X6),homepage(X5,X7)
s14(X1,X2,X3,X4,X5,X6,X7):-label(X1,X2),product(X3,X1),price(X3,X4),
      vendor(X3,X5),deliverydays(X3,X6),validto(X3,X7)
s15(X1,X2,X3,X4,X5,X6,X7,X8,X9):-product(X1,X2),price(X1,X3),vendor(X1,X4),
      label(X4,X5),country(X4,X6),publisher(X1,X4),reviewfor(X7,X2),
      reviewer(X7,X8),name(X8,X9)
```

Listing 1.4. Views s1-s10 from [9]

Consider M composed of 14 data sources defined as conjunctive views over the global schema G as in Listing 1.4; the Berlin Benchmark [8] vocabulary terms are represented as binary predicates in the conjunctive queries that define the data sources. Source $s1$ can be defined as in Listing 1.2; note that we have done just a syntactic translation from this SPARQL query to the conjunctive query presented in Listing 1.4.

For instance, *s1* retrieves information about product type, label and product feature. The `rdfs:label` predicate is a *general predicate*. Commonly, general predicates are part of the definition of many data sources, and the number of rewritings of SPARQL queries that comprise triple patterns bound to general predicates can be very large. The general predicate `rdfs:label` in query Q can be mapped to views *s1, s3-s7, s11-s15*.

```
r(X1,X2,X3,X8,X9,X10,X11,X12) :- s6(X1,X2,_0,_1,_2),
    s5(X1,_3,X3,_4,_5,_6,_7), s5(X1,_8,_9,_10,_11,X8,_12),
    s11(X1,_13,_14,_15,_16,X9,_17), s12(X1,_18,_19,_20,_21,X10,_22),
    s5(X1,_23,_24,_25,_26,_27,X11), s11(X1,_28,_29,_30,_31,_32,X12)
```

Listing 1.5. A query rewriting for Q

Listing 1.5 presents a query rewriting for Q, its subgoals cover each of the query subgoals of Q, e.g., $s6(X1, X2, \_0, \_1, \_2)$ covers the first query subgoal of Q, $label(X1, X2)$. $\psi(label(X1, X2)) = \tau(label(X1, X2))$; the mapping τ from view variables to rewriting variables is: $\tau(X1) = X1$, $\tau(X2) = X2$, $\tau(X3) = \_0$, $\tau(X4) = \_1$, $\tau(X5) = \_2$, and the mapping ψ from query variables to rewriting variables is: $\psi(Xi) = Xi$, for all Xi in the query head. Then, view $s6(X1, X2, \_0, \_1, \_2)$ is relevant for answering the first query subgoal of Q. Notice that third, fourth and fifth projected variables of $s6$ correspond to existential variables because they are not relevant to cover the first query subgoal of Q with $s6$.

To illustrate how the number of rewritings for Q can be affected by the number of data sources that use the general predicate `rdfs:label`, we run the LAV query rewriter MCDSAT [14].[3] First, if 14 data sources are considered, Q can be rewritten in 42 rewritings. For 28 data sources, there are 5,376 rewritings, and 1.12743e+10 rewritings are generated for 224 sources.[4] With one simple query, we can illustrate that the number of rewritings can be extremely large, being in the worst case exponential in the number of query subgoals and views. In addition to the problem of enumerating this large number of query rewritings, the time needed to compute them may be excessively large. Even using reasonable timeouts, only a small number of rewritings may be produced.

Table 1 shows the number of rewritings obtained by the state-of-the-art LAV rewriters GQR [4], MCDSAT [14] and MiniCon [15], when 224 views are considered for Q and timeouts are set up to 5, 10 and 20 min. Note that all these rewriters are able to produce only empty results or a small number of rewritings.

In summary, even if the LAV approach constitutes a flexible approach to integrate data from heterogeneous data sources, query rewriting and processing tasks may be unfeasible in the context of SPARQL queries. Either the number of query rewritings is too large to be enumerated or executed in a reasonable time.

---

[3] MCDSAT [14] is the only query rewriting tool publicly available that counts the number of rewritings without enumerating all of them.

[4] The 14 data sources setup is defined as in Listing 1.4, the one with 28 data sources has two views for each of the views in Listing 1.4, and the one with 224 sources has 16 views for each of the views in Listing 1.4

Table 1. Number of rewritings obtained from the rewriters GQR, MCDSAT and Mini-Con with timeouts of 5, 10 and 20 min. Using 224 views and query Q

Rewriter	5 min	10 min	20 min
GQR	0	0	0
MCDSAT	211,125	440,308	898,766
MiniCon	0	0	0

To overcome these limitations and make feasible the LAV approach for SPARQL queries, we propose a novel approach named SemLAV. SemLAV identifies and ranks the relevant views of a query, and executes the query over the data collected from the relevant views; thus, SemLAV is able to output a high proportion of the answer in a short time.

3   The SemLAV Approach

SemLAV is a scalable LAV-based approach for processing SPARQL queries. It is able to produce answers even for SPARQL queries against large integration systems with no statistics. SemLAV follows the traditional mediator-wrapper architecture [10]. Schemas exposed by the mediators and wrappers are expressed as RDF vocabularies. Given a SPARQL query Q over a global schema G and a set of sound views $M = \{v_1, \ldots, v_m\}$, SemLAV executes the original query Q rather than generating and executing rewritings as in traditional LAV approaches. SemLAV builds an instance of the global schema on-the-fly with data collected from the relevant views. The relevant views are considered in an order that enables to produce results as soon as the query Q is executed against this instance.

Contrary to traditional wrappers which populate structures that represent the heads of the corresponding views, SemLAV wrappers return RDF Graphs composed of the triples that match the triple patterns in the definition of the views. SemLAV wrappers could be more expensive in space than the traditional ones. However, they ensure that original queries are executable even for full SPARQL queries and they make query execution dependent on the number of views rather than on the number of rewritings.

To illustrate the SemLAV approach, consider a SPARQL query Q with four subgoals:

```
SELECT *
WHERE {
    ?Offer bsbm:vendor ?Vendor .
    ?Vendor rdfs:label ?Label .
    ?Offer bsbm:product ?Product .
    ?Product bsbm:productFeature ?ProductFeature .
}
```

and a set M of five views:

```
v1(P,L,T,F):-label(P,L),type(P,T),productfeature(P,F)
v2(P,R,L,B,F):-producer(P,R),label(R,L),publisher(P,B),productfeature(P,F)
v3(P,L,O,R,V):-label(P,L),product(O,P),price(O,R),vendor(O,V)
v4(P,O,R,V,L,U,H):-product(O,P),price(O,R),vendor(O,V),label(V,L),
    offerwebpage(O,U),homepage(V,H)
v5(O,V,L,C):-vendor(O,V),label(V,L),country(V,C)
```

In the traditional LAV approach, 60 rewritings are generated and the execution of all these rewritings will produce all possible answers. However, this is time-consuming and uses a non-negligible amount of memory to store data collected from views present in the rewritings. In case there are not enough resources to execute all these rewritings, as many rewritings as possible would be executed. We apply a similar idea in SemLAV, if it is not possible to consider the whole global schema instance to ensure a complete answer, then a partial instance will be built. The partial instance will include data collected from as many relevant views as the available resources allow.

The execution of the query over this partial schema instance will cover the results of executing a number of rewritings. The number of rewritings covered by the execution of Q over the partial schema instance could be exponential in the number of views included in the instance. Therefore, the size of the set of covered rewritings may be even greater than the number of rewritings executable in the same amount of time.

The order in which views are included in the partial global schema instance impacts the number of covered rewritings. Consider two different orders for including the views of the above example: v5, v1, v3, v2, v4 and v4, v2, v3, v1, v5. Table 2 considers partial global schema instances of different sizes. For each partial global schema instance, the included views and the number of covered rewritings are presented. Executing Q over the growing instances corresponds to the execution of a quite different number of rewritings. For instance, if only four views could be included with the available resources, one order corresponds to the execution of 32 rewritings while the another one corresponds to the execution of only eight rewritings. If all relevant views for query Q could be included, then a complete answer will be produced. However, the number of

Table 2. Impact of the different views ordering on the number of covered rewritings

# Included views (k)	Order one		Order two	
	Included views (V_k)	# Covered rewritings	Included views (V_k)	# Covered rewritings
1	v5	0	v4	0
2	v5, v1	0	v4, v2	2
3	v5, v1, v3	6	v4, v2, v3	12
4	v5, v1, v3, v2	8	v4, v2, v3, v1	32
5	v5, v1, v3, v2, v4	60	v4, v2, v3, v1, v5	60

relevant views could be considerably large, therefore, if we only have resources to consider k relevant views, V_k, we should consider the ones that increase the chances of obtaining answers. With no knowledge about data distribution, we can only suppose that each rewriting has nearly the same chances of producing answers. Thus, the chances of obtaining answers are proportional to the number of rewritings covered by the execution of Q over an instance that includes views in V_k.

Maximal Coverage Problem (MaxCov). Given an integer $k > 0$, a query Q on a global schema G, a set M of sound views over G, and a set R of conjunctive queries whose union is a maximally-contained rewriting of Q in M. The Maximal Coverage Problem is to find a subset V_k of M comprised of k relevant views for Q, $V_k \subseteq M \land (\forall v : v \in V_k : v \in RV(Q, M)) \land |V_k| = k$, such that the set of rewritings covered by V_k, $Coverage(V_k, R)$, is maximal for all subsets of M of size k, i.e., there is no other set of k views that can cover more rewritings than V_k. $Coverage(V_k, R)$ is defined as:

$$Coverage(V_k, R) = \{r : r \in R \land (\forall p : p \in body(r) : p \in V_k)\} \qquad (1)$$

The MaxCov problem has as an input a solution to the Maximally-Contained Rewriting problem. Nevertheless, using this for building a MaxCov solution would be unreasonable since it makes the MaxCov solution at least as expensive as the rewriting generation. Instead of generating the rewritings, we define a formula that estimates the number of covered rewritings when Q is executed over a global schema instance that includes a set of views. It is the product of the number of ways each query subgoal can be covered by the set of views. For a query $Q(\bar{X}) \coloneq p_1(\bar{X}_1), \dots p_n(\bar{X}_n)$ using only views in V_k this formula is expressed as:

$$NumberOfCoveredRewritings(Q, V_k) = \Pi_{1 \leq i \leq n} |Use(V_k, p_i(\bar{X}_i))|, \qquad (2)$$

where $Use(V_k, p) = \Sigma_{v \in V_k} \Sigma_{w \in body(v) \land covers(w,p)} 1$. This formula computes the exact number of covered rewritings when all the view variables are distinguished; this is because the coverage of each query subgoal by a given view can be considered in isolation. Otherwise, this expression corresponds to an upper bound of the number of covered rewritings of Q with respect to V_k.

Consider the second proposed ordering of the views in the above example, the numbers of views in V_4 that cover each query subgoal are:

- two for the first query subgoal (v4 and v3),
- four for the second query subgoal (v4, v2, v3 and v1),
- two for the third query subgoal (v4 and v3), and
- two for the fourth query subgoal (v2 and v1).

Thus, the number of covered rewritings is 32 ($2 \times 4 \times 2 \times 2$).

Next, we detail a solution to the MaxCov problem under the assumption that views only contain distinguished variables.

3.1   The SemLAV Relevant View Selection and Ranking Algorithm

The relevant view selection and ranking algorithm finds the views that cover each subgoal of a query. This algorithm creates a bucket for each query subgoal q, where a bucket is a set of relevant views; this resembles the first step of the Bucket algorithm [11]. Additionally, the algorithm sorts the buckets views according to the number of covered subgoals. Hence, the views that are more likely to contribute to the answer will be considered first. This algorithm is defined in Algorithm 1.

Algorithm 1 The Relevant View Selection and Ranking

Input: Q : SPARQL Query; M: Set of Views defined as conjunctive queries
Output: *Buckets*: Predicate → List<View>
  for all $q \in body(Q)$ **do**
    $buckets(q) \leftarrow \emptyset$
  end for
  for all $q \in body(Q)$ **do**
    $b \leftarrow buckets(q)$
    for all $v \in M$ **do**
      for all $w \in body(v)$ **do**
        if There are mappings τ, ψ, such that $\psi(q) = \tau(w)$ **then**
          $v_i \leftarrow \lambda(v)$ {$\lambda(v)$ replaces all variables a_i in the head of v by $\tau(a_i)$}
          $insert(b, vi)$ {add v_i to the bucket if it is not redundant}
        end if
      end for
    end for
  end for
  for all $q \in body(Q)$ **do**
    $b \leftarrow buckets(q)$
    sortBucket(buckets,b) {MergeSort with key (#covered buckets,#views subgoals)}
  end for

The mapping τ relates view variables to query variables as stated in Definition 9.

The *sortBucket(buckets, b, q)* procedure decreasingly sorts the views of bucket b according to the number of covered subgoals. Views covering the same number of subgoals are sorted decreasingly according to their number of subgoals. Intuitively, this second sort criterion prioritizes the more selective views, reducing the size of the global schema instance. The sorting is implemented as a classical *MergeSort* algorithm with a complexity of $O(|M| \times log(|M|))$.

Proposition 1. *The complexity of Algorithm 1 is $Max(O(N \times |M| \times P), O(N \times |M| \times log(|M|)))$ where N is the number of query subgoals, M is the set of views and P is the maximal number of view subgoals.*

To illustrate Algorithm 1, consider the SPARQL query Q and the previously defined views v1-v5.

Algorithm 1 creates a bucket for each subgoal in Q as shown in Table 3a.

For instance, the bucket of subgoal $vendor(O, V)$ contains v3, v4 and v5: all the views having a subgoal covering $vendor(O, V)$. The final output after executing the *sortBucket* procedure is described in Table 3b.

Views v3 and v4 cover three subgoals, but since v4 definition has more subgoals, i.e., it is more selective, v4 is placed before v3 in all the buckets.

Table 3. For query Q, buckets produced by Algorithm 1 when k views have been included. V_k is obtained by Algorithm 2 and the number of covered rewritings

(a) Unsorted buckets

vendor(O,V)	label(V,L)	product(O,P)	productfeature(P,F)
v3(P,L,O,R,V)	v1(P,L,T,F)	v3(P,L,O,R,V)	v1(P,L,T,F)
v4(P,O,R,V,L,U,H)	v2(P,R,L,B,F)	v4(P,O,R,V,L,U,H)	v2(P,R,L,B,F)
v5(O,V,L,C)	v3(P,L,O,R,V)		
	v4(P,O,R,V,L,U,H)		
	v5(O,V,L,C)		

(b) Sorted buckets

vendor(O,V)	label(V,L)	product(O,P)	productfeature(P,F)
v4(P,O,R,V,L,U,H)	v4(P,O,R,V,L,U,H)	v4(P,O,R,V,L,U,H)	v2(P,R,L,B,F)
v3(P,L,O,R,V)	v3(P,L,O,R,V)	v3(P,L,O,R,V)	v1(P,L,T,F)
v5(O,V,L,C)	v2(P,R,L,B,F)		
	v1(P,L,T,F)		
	v5(O,V,L,C)		

(c) Included views

# Included views (k)	Included views (V_k)	# Covered rewritings
1	v4	$1 \times 1 \times 1 \times 0 = 0$
2	v4, v2	$1 \times 2 \times 1 \times 1 = 2$
3	v4, v2, v3	$2 \times 3 \times 2 \times 1 = 12$
4	v4, v2, v3, v1	$2 \times 4 \times 2 \times 2 = 32$
5	v4, v2, v3, v1, v5	$3 \times 5 \times 2 \times 2 = 60$

3.2   Global Schema Instance Construction and Query Execution

Each bucket is considered as a stack of views, having on the top the view that covers more query subgoals. A global schema instance is constructed as described in Algorithm 2 by iteratively popping one view from each bucket and loading its data into the instance.

Table 3c shows how the number of covered rewritings increases as views are included into the global schema instance. Each V_k in this table is a solution to the MaxCov problem, i.e., the number of covered rewritings for each V_k is maximal. There are two possible options regarding query execution. Query can be executed each time a new view is included into the schema instance and partial results will be produced incrementally; or, it can be executed after including the k views. The first option prioritizes the time for obtaining the first answer, while the second one favors the total time to receive all the answers of Q over V_k. The first option produces results as soon as possible; however, in case of non-monotonic queries, i.e., queries where partial results may not be part of the query answer, this query processing approach should not be applied. Among non-monotonic queries, there are queries with modifiers like SORT BY or constraints like a FILTER that includes the negation of a bound expression. The execution of non-monotonic queries requires all the relevant views to be included in the global schema instance in order to produce the correct results.

Algorithm 2 The Global Schema Instance Construction and Query Execution

---

Input: Q : Query
Input: $Buckets$: Predicate $\rightarrow$ List<View> {The buckets are produced by Algorithm 1}
Input: k : Int
Output: A: Set<Answer>
  $Stacks$: Predicate $\rightarrow$ Stack<View>
  V_k : Set<View>
  G : RDFGraph
  for all $p \in domain(Buckets)$ **do**
    $Stacks(p) \leftarrow toStack(Buckets(p))$
  end for
  while $(\exists p| : \neg empty(Stacks(p))) \wedge |V_k| < k$ **do**
    for all $p \in domain(Stacks) \wedge \neg empty(Stacks(p))$ **do**
      $v \leftarrow pop(Stack(p))$
      if $v \notin V_k$ **then**
        load v into G {only if is not redundant}
        $A \leftarrow A \cup exec(Q, G)$ {Option 1: Execute Q after each successful load}
        $V_k \leftarrow V_k \cup \{v\}$
      end if
    end for
  end while
  $A \leftarrow exec(Q, G)$ {Option 2: execute before exit}

Proposition 2. *Considering conjunctive queries, the time complexity of Algorithm 2 in option 1 is $O(k \times N \times I)$, while the time complexity is $O(N \times I)$ for option 2. Where k is the number of relevant views included in the instance, N the number of query subgoals, and I is the size of the constructed global schema instance.*

Proposition 3. *Algorithm 2 finds a solution to the MaxCov problem.*

Proof. By contradiction, suppose that the set V_k is not maximal in terms of the number of covered rewritings, then there is another set V_k' of size k that covers more rewritings than V_k. By construction, V_k includes the first views of each bucket, i.e., the views that cover more query subgoals. There should exist at least one view in V_k that is not in V_k', and vice-versa. Suppose w is the first view in V_k that is not in V_k' ($w \in V_k \wedge w \notin V_k'$) , v is the first one in V_k' and is not in V_k ($v \subset V_k' \wedge v \not\subset V_k$) , and w belongs to the bucket of the query subgoal q. If v covers q, then it belongs to the bucket of q. Because V_k includes the views that cover more subgoals, if v was not included in V_k is because it covers less rewritings than w; thus, the contribution of v to the number of covered rewritings is inferior to the contribution of w. This generalizes to all the views in V_k' and not in V_k; thus, the number of rewritings covered by V_k' should be less than the number of rewritings covered by V_k. If v covers another query subgoal q' and all the query subgoals are covered at least once by views in V_k; thus, Algorithm 2 should have included it before including w and v should belong to V_k.

3.3   The SemLAV Properties

Given a SPARQL query Q over a global schema G, a set M of views over G, the set RV of views in M relevant for Q, a set R of conjunctive queries whose union is a maximally-contained rewriting of Q using M, and V_k a solution to the MaxCov problem produced by SemLAV.

- *Answer Completeness:* If SemLAV executes Q over a global schema instance I that includes all the data collected from views in RV, then it produces the complete answer. SemLAV outputs the same answers as a traditional rewriting-based query processing approach:

$$\bigcup_{r \in R} r(I(M)) = Q(\bigcup_{v \in RV} I(v)). \tag{3}$$

- *Effectiveness:* the *Effectiveness* of SemLAV is proportional to the number of covered rewritings, it is defined as:

$$Effectiveness(V_k) = \frac{|Coverage(V_k, R)|}{|R|}. \tag{4}$$

For an execution constrained by time or space, V_k could be smaller than RV.
- *Execution Time depends on* $|RV|$: The load and execution time of SemLAV linearly depends on the size of the views included in the global schema instance.
- *No memory blocking:* SemLAV guarantees to obtain a complete answer when $\bigcup_{v \in RV} I(v)$ fits into memory. If not, it is necessary to divide the set RV of relevant views into several subsets RV_i, such that each subset fits into memory and for any rewriting $r \in R$ all views $v \in body(r)$ are contained in one of these subsets.

4   Experimental Evaluation

We compare the SemLAV approach with a traditional rewriting-based approach and analyze the SemLAV effectiveness, memory consumption and throughput. In order to decide which rewriting engine will be use to compare with SemLAV, we run some preliminary experiments to compare existing state-of-the-art rewriting engines. We consider GQR [4], MCDSAT [14], MiniCon [15], and SSDSAT [16]. We execute these engines for 10 min and measure execution time and the number of rewritings generated by each engine. Additionally, we use these values to compute the throughput; throughput corresponds to number of answers obtained per second. Time is expressed in seconds; the total number of rewritings is computed for each query. Table 5 reports on all these metrics. The GQR performance is very good when the number of query rewritings is low, and it outperforms all the other engines. It also performs pretty well when the number of query rewritings is relatively low and views can cover more than a query subgoal. That is, this situation allows to speeds up the preprocessing time consumed by GQR to

Table 4. Queries and their answer size, number of subgoals, and views size

(a) Query information

Query	Answer size	# Subgoals
Q1	6.68E+07	5
Q2	5.99E+05	12
Q4	2.87E+02	2
Q5	5.64E+05	4
Q6	1.97E+05	3
Q8	5.64E+05	3
Q9	2.82E+04	1
Q10	2.99E+06	3
Q11	2.99E+06	2
Q12	5.99E+05	4
Q13	5.99E+05	2
Q14	5.64E+05	3
Q15	2.82E+05	5
Q16	2.82E+05	3
Q17	1.97E+05	2
Q18	5.64E+05	4

(b) Views size

Views	Size
V1-V34	201,250
V35-V68	153,523
V69-V102	53,370
V103-V136	26,572
V137-V170	5,402
V171-V204	66,047
V205-V238	40,146
V239-V272	113,756
V273-V306	24,891
V307-V340	11,594
V341-V374	5,402
V375-V408	5,402
V409-V442	78,594
V443-V476	99,237
V477-V510	1,087,281

build the structures required to generate the query rewritings. The MCDSAT performance is good in a larger number of queries; it can produce rewritings for more queries than the other engines, particularly in queries which a large number of triple patterns and in presence of general predicates. However, MCD-SAT does not outperform the others engines when they are able to produce the rewritings. This is because, there is an overhead in translating the problem into a logical theory which is solved using a SAT solver. The MiniCon performance is pretty good in general, but it only produces query rewritings when the space

Table 5. Comparison of state-of-the-art LAV rewriting engines for 16 queries without existential variables and nine (plus five) views defined in [9]. The five additional views allows to cover all the queries subgoals

Query	Metric	GQR	MCDSAT	MiniCon	SSDSAT	Total number of rewritings
Q1	Execution time (s)	600.00	600.00	600.00	600.00	
	Number of rewritings	0	247,304	0	0	2.04E+10
	Throughput (answers/s)	0.00	412.17	0.00	0.00	
Q2	Execution time (s)	600.00	600.00	600.00	600.00	
	Number of rewritings	0	0	0	0	1.57E+24
	Throughput (answers/s)	0.00	0.00	0.00	0.00	
Q4	Execution time (s)	25.71	84.20	5.47	600.00	
	Number of rewritings	16,184	16,184	16,184	0	1.62E+04
	Throughput (answers/s)	629.38	192.22	2,957.60	0.00	
Q5	Execution time (s)	600.00	600.00	600.00	600.00	
	Number of rewritings	0	513,629	0	0	7.48E+07
	Throughput (answers/s)	0.00	856.05	0.00	0.00	
Q6	Execution time (s)	600.00	251.51	430.10	600.00	
	Number of rewritings	0	314,432	314,432	0	3.14E+05
	Throughput (answers/s)	0.00	1,250.18	731.07	0.00	
Q8	Execution time (s)	555.49	191.69	142.63	600.00	
	Number of rewritings	157,216	157,216	157,216	0	1.57E+05
	Throughput (answers/s)	283.02	820.16	1,102.30	0.00	
Q9	Execution time (s)	0.88	32.24	0.34	49.83	
	Number of rewritings	34	34	34	34	3.40E+01
	Throughput (answers/s)	38.51	1.05	101.49	0.68	
Q10	Execution time (s)	600.00	600.00	600.00	600.00	
	Number of rewritings	0	656,140	0	0	4.40E+06
	Throughput (answers/sec)	0.00	1,093.57	0.00	0.00	
Q11	Execution time (s)	12.99	67.03	2.06	600.00	
	Number of rewritings	9,248	9,248	9,248	0	9.25E+03
	Throughput (answers/s)	712.15	137.96	4,487.14	0.00	
Q12	Execution time (s)	600.00	600.00	600.00	600.00	
	Number of rewritings	0	440,059	0	0	1.50E+09
	Throughput (answers/s)	0.00	733.43	0.00	0.00	
Q13	Execution time (s)	600.00	98.43	22.40	600.00	
	Number of rewritings	0	64,736	64,736	0	6.47E+04
	Throughput (answers/s)	0.00	657.69	2,890.52	0.00	
Q14	Execution time (s)	600.00	600.00	600.00	600.00	
	Number of rewritings	0	913,807	0	0	2.52E+06
	Throughput (answers/s)	0.00	1,523.01	0.00	0.00	
Q15	Execution time (s)	600.00	600.00	600.00	600.00	
	Number of rewritings	0	308,903	0	0	2.04E+10
	Throughput (answers/s)	0.00	514.84	0.00	0.00	
Q16	Execution time (s)	600.00	233.47	380.81	600.00	
	Number of rewritings	0	314,432	314,432	0	3.14E+05
	Throughput (answers/s)	0.00	1,346.81	825.68	0.00	

continued

Table 5. continued

Query	Metric	GQR	MCDSAT	MiniCon	SSDSAT	Total number of rewritings
Q17	Execution time (s)	3.97	67.25	1.29	600.00	
	Number of rewritings	4,624	4,624	4,624	0	4.62E+03
	Throughput (answers/s)	1,165.62	68.76	3,576.18	0.00	
Q18	Execution time (s)	600.00	600.00	600.00	600.00	
	Number of rewritings	0	463,754	0	0	1.20E+09
	Throughput (answers/s)	0.00	772.92	0.00	0.00	

of rewritings is relatively small. Finally, SSDSAT is able to handle constants; however, this feature severely impacts its performance, being able to produce rewritings only for simple cases.

4.1 Hypothesis of Our Experimentations

The hypotheses of our experimentation are:

- SemLAV loads the more relevant views of a query first, the SemLAV effectiveness should be considerably high and should produce more answers than the rest of the engines in the same amount of time.
- SemLAV builds a global schema instance using data collected from the relevant views, SemLAV may consume more space than a traditional rewriting-based approach.
- SemLAV produces results incrementally, it is able to produce answers sooner than a traditional rewriting-based approach.

4.2 Experimental Configuration

The Berlin SPARQL Benchmark (BSBM) [8] is used to generate a dataset of 10,000,736 triples using a scale factor of 28,211 products. Additionally, third-party queries and views are used to provide an unbiased evaluation of our approach. In our experiments, the goal is to study SemLAV as a solution to the MaxCov problem, and we compute the number of rewritings generated by three state-of-the-art query rewriters. From the 18 queries and 10 views defined in [9], we leave out the ones using constants (literals) because the state-of-the-art query rewriters are unable to handle constants either in the query or in the views. In total, we use 16 out of 18 queries and nine out of 10 the defined views. The query triple patterns can be grouped into chained connected star-shaped sub-queries, that have between one and twelve subgoals with only distinguished variables, i.e., queries are free of existential variable. We define five additional views to cover all the predicates in the queries. From these 14 views, we produce 476 views by horizontally partitioning each original view into 34 parts, such that each part produces 1/34 of the answers given by the original view.

Table 6. The SemLAV Effectiveness. For 10 min of execution, we report the number of relevant views included in the global schema instance, the number of covered rewritings and the achieved effectiveness. Also values for total number of views and rewritings are shown

Query	Included views	# Relevant views	# Covered rewritings	# Rewritings	Effectiveness
Q1	30	408	2.28E+06	2.04E+10	0.000112
Q2	194	408	2.05E+23	1.57E+24	0.130135
Q4	156	374	8.77E+03	1.62E+04	**0.542017**
Q5	52	374	3.13E+06	7.48E+07	0.041770
Q6	44	136	2.13E+04	3.14E+05	0.067728
Q8	81	136	9.36E+04	1.57E+05	**0.595588**
Q9	34	34	3.40E+01	3.40E+01	**1.000000**
Q10	88	408	3.20E+05	4.40E+06	0.072766
Q11	77	136	5.24E+03	9.25E+03	**0.566176**
Q12	238	408	7.70E+08	1.50E+09	**0.514286**
Q13	245	408	4.26E+04	6.47E+04	**0.657563**
Q14	46	272	1.22E+04	2.52E+06	0.004837
Q15	70	442	5.12E+08	2.04E+10	0.025144
Q16	82	136	1.90E+05	3.14E+05	**0.602941**
Q17	56	136	1.90E+03	4.62E+03	0.411765
Q18	23	374	2.80E+05	1.20E+09	0.000234

Queries and views are described in Table 4a and 4b. The size of the complete answer is computed by including all the views into an RDF-Store (Jena) and executing the queries against this centralized RDF dataset. Query definitions are included in Appendix A.

We implement wrappers as simple file readers. For executing rewritings, we use one named graph per subgoal as done in [17]. The Jena 2.7.4[5] library with main memory setup is used to store and query the graphs. The SemLAV algorithms are implemented in Java, using different threads for bucket construction, view inclusion and query execution to improve performance. The implementation is available in the project website[6].

4.3   Experimental Results

The analysis of our results focus on three main aspects: the SemLAV effectiveness, memory consumption and throughput.

To demonstrate the SemLAV effectiveness, we execute SemLAV with a timeout of 10 min. During this execution, the SemLAV algorithms select and include a subset of the relevant views; this set corresponds to V_k as a solution to the MaxCov problem. Then, we use these views to compute the number of covered rewritings using the formula given in Sect. 3. Table 6 shows the number of relevant views considered by SemLAV, the covered rewritings and the achieved effectiveness. Effectiveness is greater than or equal to 0.5 (out of 1) for almost

---

[5] http://jena.apache.org/
[6] https://sites.google.com/site/semanticlav/

Table 7. Execution of Queries Q1, Q2, Q4-Q6, Q8-Q18 using SemLAV, MCDSAT, GQR and MiniCon, using 20 GB of RAM and a timeout of 10 min. It is reported the number of answers obtained, wrapper time (WT), graph creation time (GCT), plan execution time (PET), total time (TT), time of first answer (TFA), number of times original query is executed (#EQ), maximal graph size (MGS) in terms of number of triples and throughput (number of answers obtained per millisecond)

Query	Approach	Answer Size	%	WT	GCT	PET	TT	TFA	#EQ	MGS	Throughput (answers/ms)
Q1	SemLAV	22,660,216	33	45,434	8,322	547,310	606,697	**6,370**	15	810,638	**37.3501**
	MCDSAT	290	0	13,688	202	299,546	609,381	309,952		810,409	0.0005
	GQR	0	0	0	0	0	600,415	>600,000		0	0.0000
	MiniCon	0	0	0	0	0	600,136	>600,000		0	0.0000
Q2	SemLAV	590,000	98	177,020	30,676	392,439	600,656	**260,333**	66	1,040,373	**0.9823**
	MCDSAT	0	0	15,519	105	7,058	681,246	>600,000		848,276	0.0000
	GQR	0	0	0	0	0	654,483	>600,000		0	0.0000
	MiniCon	0	0	0	0	0	600,054	>600,000		0	0.0000
Q4	SemLAV	287	100	555,528	73,771	327	660,938	**104,501**	47	3,659,707	**0.0004**
	MCDSAT	0	0	154,451	371	181,387	601,590	>600,000		279,896	0.0000
	GQR	0	0	557,125	1,181	11,784	600,665	>600,000		84,046	0.0000
	MiniCon	0	0	413,871	650	91,136	601,750	>600,000		177,838	0.0000
Q5	SemLAV	564,220	100	523,084	65,333	44,102	632,809	**116,037**	28	3,396,134	**0.8916**
	MCDSAT	0	0	398,517	384	26,287	601,731	>600,000		424,431	0.0000
	GQR	0	0	0	0	0	600,481	>600,000		0	0.0000
	MiniCon	0	0	0	0	0	600,132	>600,000		0	0.0000
Q6	SemLAV	118,258	59	547,763	62,896	13,291	625,173	**43,306**	24	2,931,316	**0.1892**
	MCDSAT	5,776	2	401,026	1,029	55,684	601,678	105,752		91,900	0.0096
	GQR	0	0	0	0	0	600,510	>600,000		0	0.0000
	MiniCon	3,697	1	193,817	248	51,300	637,514	418,169		2,184,680	0.0058
Q8	SemLAV	564,220	100	428,745	66,383	132,373	627,612	**5,393**	42	4,489,016	**0.8990**
	MCDSAT	16,595	2	403,133	576	65,935	603,297	113,211		256,382	0.0275
	GQR	1,706	0	330,065	194	31,587	607,594	272,737		1,264,385	0.0028
	MiniCon	467	0	198,384	349	271,398	616,114	166,776		1,265,295	0.0008
Q9	SemLAV	28,211	100	2,938	697	1,338	5,107	**1,235**	18	169,839	**5.5240**
	MCDSAT	28,211	100	5,609	445	1,643	41,505	34,392		5,417	0.6797
	GQR	28,211	100	3,310	132	1,281	5,709	1,435		5,417	4.9415
	MiniCon	28,211	100	3,086	129	1,362	5,004	862		5,417	5.6377
Q10	SemLAV	2,993,175	100	161,047	25,659	417,234	607,841	**9,810**	44	869,340	**4.9243**
	MCDSAT	332,488	11	19,801	67	383,421	600,000	207,191		603,769	0.5541
	GQR	0	0	0	0	0	600,639	>600,000		0	0.0000
	MiniCon	0	0	0	0	0	600,138	>600,000		0	0.0000
Q11	SemLAV	2,993,175	100	195,950	27,442	377,255	601,042	**8,352**	43	816,308	**4.9800**
	MCDSAT	1,943,141	64	141,876	389	391,852	600,000	72,939		402,528	3.2386
	GQR	1,442,134	48	248,275	689	340,937	600,000	14,435		307,089	2.4036
	MiniCon	1,956,539	65	217,321	415	385,019	605,021	6,832		402,539	3.2338
Q12	SemLAV	598,635	100	258,097	41,062	303,023	609,509	**5,784**	121	1,041,369	**0.9822**
	MCDSAT	0	0	424,369	498	15,271	607,408	>600,000		509,271	0.0000
	GQR	0	0	0	0	0	600,418	>600,000		0	0.0000
	MiniCon	0	0	0	0	0	600,189	>600,000		0	0.0000
Q13	SemLAV	598,635	100	452,288	65,043	126,345	671,893	**183,844**	124	3,509,975	**0.8910**
	MCDSAT	0	0	250,542	312	141,728	610,452	>600,000		402,531	0.0000
	GQR	0	0	36,563	344	19,757	600,376	>600,000		31,948	0.0000
	MiniCon	0	0	143,879	625	219,882	605,727	>600,000		206,689	0.0000
Q14	SemLAV	344,885	61	544,919	58,563	32,752	636,387	**29,201**	24	2,921,646	**0.5419**
	MCDSAT	10,308	1	382,674	587	63,689	614,123	133,200		1,206,075	0.0168
	GQR	0	0	0	0	0	600,714	>600,000		0	0.0000
	MiniCon	0	0	0	0	0	600,319	>600,000		0	0.0000
Q15	SemLAV	282,110	100	471,609	63,548	109,762	645,172	**2,911**	37	3,255,223	**0.4373**
	MCDSAT	8,298	2	90,061	271	168,041	622,474	217,445		361,882	0.0133
	GQR	0	0	0	0	0	819,679	>600,000		0	0.0000
	MiniCon	0	0	0	0	0	600,171	>600,000		0	0.0000
Q16	SemLAV	282,110	100	407,107	53,611	187,986	648,826	**2,531**	46	3,356,755	**0.4348**
	MCDSAT	8,298	2	437,590	852	32,015	601,584	103,641		74,682	0.0138
	GQR	1	0	26,460	79	94	619,761	619,702		1,136,305	0.0000
	MiniCon	252	0	110,366	181	122,022	603,821	400,416		1,151,769	0.0004

Table 7. continued

Query	Approach	Answer		Time (ms)					#EQ	MGS	Throughput
		Size	%	WT	GCT	PET	TT	TFA			(answers/ms)
Q17	SemLAV	197,112	100	547,255	67,857	28,783	644,090	**1,504**	32	3,002,144	**0.3060**
	MCDSAT	156,533	79	412,525	1,727	60,858	600,067	70,476		23,192	0.2609
	GQR	45,037	22	245,953	177	350,406	600,000	27,178		1,098,117	0.0751
	MiniCon	5,779	2	262,608	361	334,810	600,001	26,952		1,099,508	0.0096
Q18	SemLAV	0	0	582,334	65,083	3,543	651,094	>600,000	12	2,806,533	0.0000
	MCDSAT	0	0	256,304	257	100,820	607,091	>600,000		411,901	0.0000
	GQR	0	0	0	0	0	600,791	>600,000		0	0.0000
	MiniCon	0	0	0	0	0	600,186	>600,000		0	0.0000

half of the queries. SemLAV maximizes the number of covered rewritings by considering views that cover more subgoals first.

The observed results confirm that the SemLAV effectiveness is considerably high. Effectiveness depends on the number of relevant views, but this number is bounded to the number of relevant views that can be stored in memory. As expected, the SemLAV approach could require more space than the traditional rewriting-based approach. SemLAV builds a global schema instance that includes all the relevant views in V_k, whereas a traditional rewriting-based approach includes only the views in one rewriting at the time. Table 7 shows the maximal graph size in both approaches. SemLAV can use up to 129 times more memory than the traditional rewriting-based approach (for Q17). SemLAV can use less memory than the traditional rewriting-based approach (for Q1) for relevant views with overlapped data.

We calculate the throughput as the number of answers divided by the total execution time. For SemLAV, this time includes view selection and ranking, contacting data sources using the wrappers, including data into the global schema instance, and query execution time. For the traditional rewriting-based approach, this time includes rewriting time, instead of view selection and ranking. Table 7 shows for each query: number of answers, execution time, number of times the query is executed and throughput. Notice that SemLAV executes the query whenever a new relevant view has been included in the global schema instance and the query execution thread is active.

The difference in the answer size and throughput is impressive, e.g., for Q1 SemLAV produces 37.3501 answers/ms, while the other approach produces up to 0.0005 answers/ms. This huge difference is caused by the differences between the complexity of the rewriting generation and the SemLAV view selection and ranking algorithm, and between the number of rewritings and number of relevant views. This makes possible to generate answers sooner. Column TFA of Table 7 shows the time for the first answer; TFA is impacted by executing the query as soon as possible, according to option 1 given in Algorithm 2. Only for query $Q18$ SemLAV does not produce any answer in 10 min. This is because the views included in the global schema instance are large (around one million triples per view) and do not contribute to the answer; consequently, almost all the execution time is spent in transferring data from the relevant views. SemLAV produces answers sooner in all the other cases. Moreover, SemLAV also achieves complete answer in 11 of 16 queries in only 10 min.

In summary, the results show that SemLAV is effective and efficient and produces more answers sooner than a traditional rewriting-based approach. SemLAV makes the LAV approach feasible for processing SPARQL queries.

5   State of the Art

In recent years, several approaches have been proposed for querying the Web of Data [18–22]. Some tools address the problem of choosing the sources that can be used to execute a query [21,22]; others have developed techniques to adapt query processing to source availability [18,21]. Finally, frameworks to retrieve and manage Linked Data have been defined [19,21], as well as strategies for decomposing SPARQL queries against federations of endpoints [6]. All these approaches assume that queries are expressed in terms of RDF vocabularies used to describe the data in the RDF sources; thus, their main challenge is to effectively select the sources, and efficiently execute the queries on the data retrieved from the selected sources. In contrast, SemLAV attempts to integrate data sources, and relies on a global schema to describe data sources and to provide a unified interface to the users. As a consequence, in addition to collecting and processing data transferred from the selected sources, SemLAV decides which of these sources need to be contacted first, to quickly answer the query.

Three main paradigms have been proposed to integrate dissimilar data sources. In GAV mediators, entities in the global schema are semantically described using views in terms of the data sources. In consequence, including or updating data sources may require the modification of a large number of mappings [3]. In contrast, the LAV approach, new data sources can be easily integrated [3]; further, data sources that publish entities of several concepts in the global schema, can be naturally defined as LAV views. Thus, the LAV approach is best suited for applications with a stable global schema but with changing data sources; contrary, the GAV approach is more suitable for applications with stable data sources and a changing global schema. Finally, a more general approach named Global-Local-As-View (GLAV) allows the definition of mappings where views on the global schema are mapped to views of the data sources. Recently, Knoblock et al. [23] and Taheriyan et al. [24] proposed Karma, a system to semi-automatically generate source descriptions as GLAV views on a given ontology. Karma makes GLAV views a solution to consume open data as well as to integrate and populate these sources into the LOD cloud.

GLAV views are suitable not only to describe sources, but also to provide the basis for the dynamic integration of open data and Web APIs into the LOD cloud. Further, theoretical results presented by Calvanese et al. [7] establish that for conjunctive queries against relational schemas, GLAV query processing techniques can be implemented as the combination of the resolution of the query processing tasks with respect to the LAV component of the GLAV views followed by query unfolding tasks on the GAV component. Thus, SemLAV can be easily extended to manage GLAV query processing tasks, and provides the basis to integrate existing GLAV views. Additionally, SemLAV can be used to develop SPARQL endpoints that dynamically access up-to-date data from the data sources or Web APIs defined by the generated GLAV views.

The problem of rewriting a query into queries on the data sources is a relevant problem in integration systems [25]. A great effort has been made to provide solutions able to produce query rewritings in the least time possible and to scale up to a large number of views. Several approaches have been defined, e.g., MCD-SAT [14], GQR [4], Bucket Algorithm [25], and MiniCon [11]. Recently, Le et al. [17] propose a solution to identify and combine GAV SPARQL views that rewrite SPARQL queries against a global vocabulary, and Izquierdo et al. [16] extend the MCDSAT rewriter with preferences to identify the combination of semantic services that rewrite a user request. Recently, Montoya et al. propose GUN [26], a strategy to maximize the number of answers obtained from a given set of k rewritings; GUN aggregates the data obtained from the relevant views present in those k rewritings and executes the query over it. Even if GUN could maximize the number of obtained answers, it would still depend on query rewritings as input, and has no criteria to order the relevant views.

We address this problem and propose SemLAV, a query processing technique for RDF store architectures that provides a uniform interface to data sources that have been defined using the LAV paradigm [27]. SemLAV gets rid of the query rewriter, and focuses on selecting relevant views for each subgoal of the query. Moreover, SemLAV decides which relevant views will be contacted first, and includes the retrieved data into a global schema instance where the query is executed. At the cost of memory consumption, SemLAV is able to quickly produce answers first, and compute a *more complete* answer when the rest of the engines fail. Since the number of valid query rewritings can be exponential in the number of views, providing an effective and efficient semantic data management technique as SemLAV is a relevant contribution to the implementation of integration systems, and provides the basis for feasible and dynamic semantic integration architectures in the Web of Data.

An alternative approach for data integration is Data Warehousing [28], where data is retrieved from the sources and stored in a repository. In this context, query optimization relies on materialized views that allows to speed up the execution time. Selecting the best set of views to be materialized is a complex problem that has been deeply studied in the literature [9, 29–32]. Commonly approaches attempt to select this set of views according to an expected workload and available resources. Recently, Castillo-Espinola [9] propose an approach where materialized views correspond to indexes for SPARQL queries that allow to speed up query execution time. Although these approaches may considerably improve performance in average, only queries that can be rewritten using the materialized views will be benefited. Further, the cost of the view maintainability process can be very high if data frequently changes and it needs to be kept up-to-date to ensure answer correctness.

SemLAV also relies on view definitions, but views are temporally included in the global schema instance during query execution; thus, data is always up-to-date. Furthermore, the number of views to be considered is not limited. The only limitation depends on the physical resources available to perform a particular query. Nevertheless, it is important to highlight that the number of relevant views for answering one query is, in the general case, considerably smaller than the total number of views in the integration system.

6 Conclusions and Future Work

In this paper, we presented SemLAV, a Local-As-View mediation technique that allows to perform SPARQL queries over views without facing problems of NP-completeness, exponential number of rewritings or restriction to conjunctive SPARQL queries. This is obtained at the price of including relevant views into a global schema instance which is space consuming. However, we demonstrated that, even if only a subset of relevant views is included, we obtain more results than traditional rewriting-based techniques. Chances of producing results are higher, if the number of covered rewritings is maximized as defined in the MaxCov problem. We proved that our ranking strategy maximizes the number of covered rewritings.

SemLAV opens a new way to execute SPARQL queries for LAV mediators that is tractable. As perspectives, the performance of SemLAV can be greatly improved by parallelizing views inclusion. Currently, SemLAV includes views sequentially due to Jena restrictions. If views were included in parallel, time to get first results would be greatly improved. Additionally, the strategy of producing results as soon as possible, can deteriorate the overall throughput. If users want to improve overall throughput, then the query should be executed once after all the views in V_k have been included. It could be also interesting to design an execution strategy where SemLAV would execute under constrained space. In this case, the problem would be to find the minimum set of relevant views that would fit in the available space and produce the maximal number of answers. All these problems will be part of our future works.

Acknowledgments. We thank C. Li for providing his MiniCon code, and J. L. Ambite and G. Konstantinidis for sharing the GQR code for the evaluation. This work is partially supported by the French National Research agency (ANR) through the KolFlow project (code: ANR-10-CONTINT-025), part of the CONTINT research program, and by USB-DID.

A Queries

In our experimental study, we evaluate the SPARQL queries proposed by Castillo-Espinola [9]. We only consider the SPARQL queries without constants or literals due to limitations of state-of-the-art rewriters.

```
SELECT *
WHERE {
    ?X1 rdfs:label ?X2 .
    ?Xr rdf:type ?X3 .
    ?X1 bsbm:productFeature ?X4 .
    ?X1 bsbm:productFeature ?X5 .
    ?X1 bsbm:productPropertyNumeric1 ?X6 .
}
```

Listing 1.6. Q1

```
SELECT *
WHERE {
    ?X1 rdfs:label ?X2 .
    ?X1 rdfs:comment ?X3 .
    ?X1 bsbm:producer ?X4 .
    ?X4 rdfs:label ?X5 .
    ?X1 dc:publisher ?X4 .
    ?X1 bsbm:productFeature ?X6 .
    ?X6 rdfs:label ?X7 .
    ?X1 bsbm:productPropertyTextual1 ?X8 .
    ?X1 bsbm:productPropertyTextual2 ?X9 .
    ?X1 bsbm:productPropertyTextual3 ?X10 .
    ?X1 bsbm:productPropertyNumeric1 ?X11 .
    ?X1 bsbm:productPropertyNumeric2 ?X12 .
}
```

Listing 1.7. Q2

```
SELECT *
WHERE {
    ?X1 rdfs:label ?X2 .
    ?X1 foaf:homepage ?X3 .
}
```

Listing 1.8. Q4

```
SELECT *
WHERE {
    ?X1 bsbm:vendor ?X2 .
    ?X1 bsbm:offerWebpage ?X3 .
    ?X2 rdfs:label ?X4 .
    ?X2 foaf:homepage ?X5 .
}
```

Listing 1.9. Q5

```
SELECT *
WHERE {
    ?X1 bsbm:reviewFor ?X2 .
    ?X1 rev:reviewer ?X3 .
    ?X1 bsbm:rating1 ?X4 .
}
```

Listing 1.10. Q6

```
SELECT *
WHERE {
    ?X1 bsbm:offerWebpage ?X2 .
    ?X1 bsbm:price ?X3 .
    ?X1 bsbm:deliveryDays ?X4 .
}
```

Listing 1.11. Q8

```
SELECT *
WHERE {
    ?X1 bsbm:productPropertyNumeric1 ?X2 .
}
```

Listing 1.12. Q9

```
SELECT *
WHERE {
    ?X1 rdfs:label ?X2 .
    ?X1 rdf:type ?X3 .
    ?X1 bsbm:productFeature ?X4 .
}
```

Listing 1.13. Q10

```
SELECT *
WHERE {
    ?X1 rdf:type ?X2 .
    ?X1 bsbm:productFeature ?X3 .
}
```

Listing 1.14. Q11

```
SELECT *
WHERE {
    ?X1 bsbm:producer ?X2 .
    ?X2 rdfs:label ?X3 .
    ?X1 dc:publisher ?X2 .
    ?X1 bsbm:productFeature ?X4 .
}
```

Listing 1.15. Q12

```
SELECT *
WHERE {
    ?X1 bsbm:productFeature ?X2 .
    ?X2 rdfs:label ?X3 .
}
```

Listing 1.16. Q13

```
SELECT *
WHERE {
    ?X1 bsbm:producer ?X2 .
    ?X3 bsbm:product ?X1 .
    ?X3 bsbm:vendor ?X4 .
}
```

Listing 1.17. Q14

```
SELECT *
WHERE {
    ?X1 rdfs:label ?X2 .
    ?X3 bsbm:reviewFor ?X1 .
    ?X3 rev:reviewer ?X4 .
    ?X4 foaf:name ?X5 .
    ?X3 dc:title ?X6 .
}
```

Listing 1.18. Q15

```
SELECT *
WHERE {
    ?X1 bsbm:reviewFor ?X2 .
    ?X1 dc:title ?X3 .
    ?X1 rev:text ?X4 .
}
```

Listing 1.19. Q16

```
SELECT *
WHERE {
    ?X1 bsbm:reviewFor ?X2 .
    ?X1 bsbm:rating1 ?X3 .
}
```

Listing 1.20. Q17

```
SELECT *
WHERE {
    ?X1 bsbm:product ?X2 .
    ?X2 rdfs:label ?X3 .
    ?X1 bsbm:vendor ?X4 .
    ?X1 bsbm:price ?X5 .
}
```

Listing 1.21. Q18

References

1. Bizer, C., Heath, T., Berners-Lee, T.: Linked data - the story so far. Int. J. Semant. Web Inf. Syst. **5**, 1–22 (2009)
2. Abiteboul, S., Manolescu, I., Rigaux, P., Rousset, M.C., Senellart, P.: Web Data Management. Cambridge University Press, New York (2011)
3. Ullman, J.D.: Information integration using logical views. Theor. Comput. Sci. **239**, 189–210 (2000)
4. Konstantinidis, G., Ambite, J.L.: Scalable query rewriting: a graph-based approach. In: Sellis, T.K., Miller, R.J., Kementsietsidis, A., Velegrakis, Y., (eds.): SIGMOD Conference, pp. 97–108. ACM (2011)
5. Vidal, M.-E., Ruckhaus, E., Lampo, T., Martínez, A., Sierra, J., Polleres, A.: Efficiently joining group patterns in SPARQL queries. In: Aroyo, L., Antoniou, G., Hyvönen, E., ten Teije, A., Stuckenschmidt, H., Cabral, L., Tudorache, T. (eds.) ESWC 2010, Part I. LNCS, vol. 6088, pp. 228–242. Springer, Heidelberg (2010)
6. Schwarte, A., Haase, P., Hose, K., Schenkel, R., Schmidt, M.: Fedx: Optimization techniques for federated query processing on linked data. In: [33] 601–616
7. Calvanese, D., Giacomo, G.D., Lenzerini, M., Vardi, M.Y.: Query processing under glav mappings for relational and graph databases. PVLDB **6**, 61–72 (2012)
8. Bizer, C., Schultz, A.: The berlin sparql benchmark. Int. J. Semant. Web Inf. Syst. **5**, 1–24 (2009)
9. Castillo-Espinola, R.: Indexing RDF data using materialized SPARQL queries. Ph.D. thesis, Humboldt-Universität zu Berlin (2012)
10. Wiederhold, G.: Mediators in the architecture of future information systems. IEEE Comput. **25**, 38–49 (1992)
11. Halevy, A.Y.: Answering queries using views: a survey. VLDB J. **10**, 270–294 (2001)
12. Lenzerini, M.: Data integration: a theoretical perspective. In: Popa, L., Abiteboul, S., Kolaitis, P.G., (eds.) PODS, pp. 233–246. ACM (2002)
13. Doan, A., Halevy, A.Y., Ives, Z.G.: Principles of Data Integration. Morgan Kaufmann, Waltham (2012)
14. Arvelo, Y., Bonet, B., Vidal, M.E.: Compilation of query-rewriting problems into tractable fragments of propositional logic. In: AAAI, pp. 225–230. AAAI Press (2006)
15. Pottinger, R., Halevy, A.Y.: Minicon: a scalable algorithm for answering queries using views. VLDB J. **10**, 182–198 (2001)
16. Izquierdo, D., Vidal, M.-E., Bonet, B.: An expressive and efficient solution to the service selection problem. In: Patel-Schneider, P.F., Pan, Y., Hitzler, P., Mika, P., Zhang, L., Pan, J.Z., Horrocks, I., Glimm, B. (eds.) ISWC 2010, Part I. LNCS, vol. 6496, pp. 386–401. Springer, Heidelberg (2010)
17. Le, W., Duan, S., Kementsietsidis, A., Li, F., Wang, M.: Rewriting queries on sparql views. In: Srinivasan, S., Ramamritham, K., Kumar, A., Ravindra, M.P., Bertino, E., Kumar, R. (eds.) WWW, pp. 655–664. ACM (2011)

18. Acosta, M., Vidal, M.E., Lampo, T., Castillo, J., Ruckhaus, E.: Anapsid: an adaptive query processing engine for sparql endpoints. In: [33] 8–34
19. Basca, C., Bernstein, A.: Avalanche: putting the spirit of the web back into semantic web querying. In: Polleres, A., Chen, H. (eds.) ISWC Posters & Demos, Volume 658 of CEUR Workshop Proceedings. http://CEUR-WS.org (2010)
20. Harth, A., Hose, K., Karnstedt, M., Polleres, A., Sattler, K.U., Umbrich, J.: Data summaries for on-demand queries over linked data. In: Rappa, M., Jones, P., Freire, J., Chakrabarti, S. (eds.): WWW, pp. 411–420. ACM (2010)
21. Hartig, O.: Zero-knowledge query planning for an iterator implementation of link traversal based query execution. [34] 154–169
22. Ladwig, G., Tran, T.: Sihjoin: querying remote and local linked data. [34] 139–153
23. Knoblock, C.A., Szekely, P.A., Ambite, J.L., Gupta, S., Goel, A., Muslea, M., Lerman, K., Mallick, P.: Interactively mapping data sources into the semantic web. In: Kauppinen, T., Pouchard, L.C., Keßler, C. (eds.): LISC, CEUR Workshop Proceedings, vol. 783, CEUR-WS.org (2011)
24. Taheriyan, M., Knoblock, C.A., Szekely, P., Ambite, J.: Rapidly integrating services into the linked data cloud. In: Cudré-Mauroux, P., et al. (eds.) ISWC 2012, Part I. LNCS, vol. 7649, pp. 559–574. Springer, Heidelberg (2012)
25. Levy, A.Y., Rajaraman, A., Ordille, J.J.: Querying heterogeneous information sources using source descriptions. In: Vijayaraman, T.M., Buchmann, A.P., Mohan, C., Sarda, N.L. (eds.): VLDB, pp. 251–262. Morgan Kaufmann (1996)
26. Montoya, G., Ibáñez, L.-D., Skaf-Molli, H., Molli, P., Vidal, M.-E.: GUN: an efficient execution strategy for querying the web of data. In: Decker, H., Lhotská, L., Link, S., Basl, J., Tjoa, A.M. (eds.) DEXA 2013, Part I. LNCS, vol. 8055, pp. 180–194. Springer, Heidelberg (2013)
27. Levy, A.Y., Mendelzon, A.O., Sagiv, Y., Srivastava, D.: Answering queries using views. In: Yannakakis, M. (ed.): PODS, pp. 95–104. ACM Press (1995)
28. Theodoratos, D., Sellis, T.K.: Data warehouse configuration. In: Jarke, M., Carey, M.J., Dittrich, K.R., Lochovsky, F.H., Loucopoulos, P., Jeusfeld, M.A. (eds.): VLDB, pp. 126–135. Morgan Kaufmann (1997)
29. Gupta, H.: Selection of views to materialize in a data warehouse. In: Afrati, F.N., Kolaitis, P.G. (eds.) ICDT 1997. LNCS, vol. 1186, pp. 98–112. Springer, Heidelberg (1997)
30. Chirkova, R., Halevy, A.Y., Suciu, D.: A formal perspective on the view selection problem. VLDB J. **11**, 216–237 (2002)
31. Karloff, H.J., Mihail, M.: On the complexity of the view-selection problem. In: Vianu, V., Papadimitriou, C.H. (eds.): PODS, pp. 167–173. ACM Press (1999)
32. Goasdoué, F., Karanasos, K., Leblay, J., Manolescu, I.: View selection in semantic web databases. PVLDB **5**, 97–108 (2011)
33. Aroyo, L., Welty, C., Alani, H., Taylor, J., Bernstein, A., Kagal, L., Noy, N.F., Blomqvist, E. (eds.): ISWC 2011, Part I. LNCS, vol. 7031. Springer, Heidelberg (2011)
34. Antoniou, G., Grobelnik, M., Simperl, E.P.B., Parsia, B., Plexousakis, D., Leenheer, P.D., Pan, J.Z. (eds.): ESWC 2011, Part I. LNCS, vol. 6643. Springer, Heidelberg (2011)

Query Reformulation in PDMS Based on Social Relevance

Angela Bonifati[1], Gianvito Summa[2]([⊠]), Esther Pacitti[3], and Fady Draidi[4]

[1] University of Lille 1, Cité Scientifique, Lille, France
angela.bonifati@gmail.com
[2] University of Basilicata, Viale dell'Ateneo Lucano, Potenza, Italy
gianvito.summa@gmail.com
[3] LIRMM - University of Montpellier II, Rue Ada, Montpellier, France
esther.pacitti@lirmm.fr
[4] An Najah National University, Nablus, Palestine
draif@najah.edu

Abstract. We consider peer-to-peer data management systems (PDMS), where each peer maintains mappings between its schema and some acquaintances, along with social links with peer friends. In this context, we deal with reformulating conjunctive queries from a peer's schema into other peer's schemas. Precisely, queries against a peer node are rewritten into queries against other nodes using schema mappings thus obtaining query rewritings. Unfortunately, not all the obtained rewritings are relevant to a given query, as the information gain may be negligible or the peer is not worth exploring. On the other hand, the existence of social links with peer friends might be useful to get relevant rewritings. Therefore, we propose a new notion of 'relevance' of a query with respect to a mapping that encompasses both a local relevance (the relevance of the query w.r.t. the mapping) and a global relevance (the relevance of the query w.r.t. the entire network). Based on this notion, we have conceived a new query reformulation approach for social PDMS which achieves great accuracy and flexibility. To this purpose, we combine several techniques: (i) social links are expressed as FOAF (Friend of a Friend) links to characterize peer's friendship; (ii) concise mapping summaries are used to obtain mapping descriptions; (iii) local semantic views (LSV) are special views that contain information about mappings captured from the network by using gossiping techniques. Our experimental evaluation, based on a prototype on top of PeerSim and a simulated network demonstrate that our solution yields greater recall, compared to traditional query translation approaches proposed in the literature.

Keywords: Query reformulation · Mapping evaluation · Social PDMS · AF-IMF measure

1 Introduction

In the last decade, we have witnessed a dramatic shift in the scale of distributed and heterogeneous databases [7,17]: they have become larger, more dispersed

A. Hameurlain et al. (Eds.): TLDKS XIII, LNCS 8420, pp. 59–90, 2014.
DOI: 10.1007/978-3-642-54426-2_3, © Springer-Verlag Berlin Heidelberg 2014

and semantically interconnected networks of peers, exhibiting varied schemas and instances. A P2P data management system (PDMS) [16] is an ad-hoc collection of independent peers that have formed a network in order to map and share their data. For example, consider an online scientific community[1], that uses an underlying P2P infrastructure for data sharing. In particular, each peer embodies a medical doctor or a physician, who enters the community to share her clinical data (yet hiding sensitive patient record data) with a subset of her colleagues (for instance, she may share data about clinical trials, diseases, treatments, patients' histories, drug doses, opportunity of grants and so on). In such a way, for example, if she has doubts about a treatment or a patient's history, she may consult her previous cases by querying her data (the ones she knows) and, if no useful results are given as output, she could query her colleagues' data that, unfortunatly, have been structured according to their own peer's schema.

In fact, peers in such example typically have heterogeneous schemas, with no mediated or centralized schema. Still, to process a query over the PDMS, the data needs to be translated from one peer's schema to another peer's schema. To address this problem, PDMS maintains a set of mappings or correspondences between a peer schema and a sufficiently small number of other peer schemas, called *acquaintances*. The mappings between the local schema and the acquaintance schema can be manually provided, or, alternatively, computed via an external schema matching tool [9,26,29].

Each doctor likes to exchange specific data about treatments and patients with the peers she trusts, and/or she is friend with. Additionally, she may not find the information within her set of acquaintances, and may need to look for colleagues she has never met before.

In order to cope with data heterogeneity in PDMS, queries are formulated against a local peer schema, and translated against each schema of the peer acquaintances, and transitively on: from the schema of peer acquaintances towards the schema of the acquaintances of the acquaintances. Notice that the translation can be done according or against the mapping direction, as detailed in the following section.

This problem, called *query reformulation*, has been addressed in the literature by schema mappings tools [6,26,29], and proved to be effective in PDMS [18]. However, a fundamental limitation of the above tools is the fact that query translation is essentially enacted on every peer by tracking all the mappings, whereas in a realistic scenario, only semantically relevant mappings must be exploited[2]. E.g. in our online community, each doctor would like to exchange specific data about treatments and patients only with the peers that provide relevant information (members of the same lab or former university mates), rather than with every peer. Similarly, she may be willing to know who else,

---

[1] This example has been inspired by a web-based online trusted physician network, https://www.ozmosis.com/home, 'where good doctors go to become great doctors'.

[2] Notice that the *relevance* of a mapping is useful to discriminate the importance of such mapping wrt other mappings. Indeed, the relevance is a criterion to rank mappings in order to choose the best ones towards which the query has to be translated.

among the doctors in her community, or among her friend doctors, has worked on similar cases.

As the above example (typical of professional social networking) suggests, social relationships (or friendships) between community members are also crucial to locate relevant information. Similarly, in order to identify relevant mappings, we exploit friendship links between peers, in addition to acquaintances, in what we call social PDMS[3]. As in social networks, by establishing a friendship link, a peer p_i can become friend with a peer p_j and share peer information. In our case, the peer information we are interested in is local semantic mappings, i.e. the mappings of that peer towards its acquaintances. They express meaningful semantic relationships between elements in heterogeneous schemas on different peers. In addition, to capture peer friendship, we adopt the Friend of a Friend (FOAF) vocabularies [4]. FOAF is a way to provide a detailed description of users, i.e. peers in our context, and their relationships using RDF syntax. We have adapted the FOAF files to PDMS and extended the FOAF syntax to also point to the mapping summaries of a peer's friends.

In this paper, we tackle the problem of query reformulation for conjunctive queries in social PDMS. Based on a new notion of *relevance* of a query with respect to a mapping, we propose a query reformulation approach, using both semantic mappings and friendship links, thus biasing the query translation only towards *relevant* peers, according to a novel relevance metric.

To precisely define the notion of relevance of a query with respect to a mapping, we propose a novel metric called AF-IMF measure, which takes into account the semantic proximity between the query and the mappings which have to be taken into account. However, the above metric would need to be computed distributively, and to do so, would have to theoretically contact every peer in the network. To address this difficulty, we store on each peer a *local semantic view (LSV)*, that offers a synthetic description of the mapping components of external peers in the network. To feed and keep updated such views, we adopt

These techniques refer to the probabilistic exchange of mappings between two peers, thus leading to the endless process of making two random peers communicate among each other. We adapt gossiping to our context by periodically refreshing the local semantic view on each peer, based on gossiped atoms; by means of such semantic views, promising semantic paths can be undertaken in the network, such that, for a given query, the most relevant mappings can be located and/or the most relevant peer friends can be reached.

Contributions. To sum up, the main contributions of this paper are threefold:

(i) We propose a novel notion of relevance of query with respect to a mapping, along with that of a relevant rewriting; we characterize each mapping in the entire collection of mappings present in the network with a new metric, the AF-IMF measure, which precisely identifies the most interesting mappings, towards which query translation should be directed.

---

[3] Notice that a friendship link between two peers is a symmetric relationship and does not imply that such peers have to be *acquaintances* with each other.

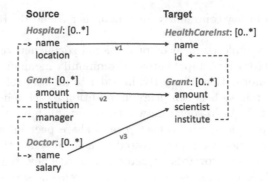

Fig. 1. A Schema Mapping Example

(ii) We propose algorithms that, given an input query Q, and a set of mappings between peers schemas, do the following: translate the query into Q^t only against the relevant mappings by adopting our new evaluation metric; exploit friendship links among peers to possible enlarge the set of mappings and bias the search towards interesting peers; exploit semantic gossiping to discover new relevant mappings and friends, thus increasing the number of query rewritings. To the best of our knowledge, these algorithms advance the state of art of query reformulation in PDMS (more details in Sect. 6).

(iii) We provide an extensive experimental evaluation by running our algorithms on a simulated network built on top of PeerSim, which demonstrates that our solution yields greater recall, compared to traditional query translation approaches.

The paper is organized as follows. Section 2 presents the background and the problem definition. Section 3 introduces our framework, while Sect. 4 and Sect. 5 describe our algorithms and the experimental assessment that has been conducted. Finally, Sect. 6 discusses the related work and Sect. 7 concludes the paper.

2   Problem Definition

In this section, we first present the background of schema mappings and P2P networks, and then we detail the problem statement.

2.1   Schema Mapping Model

Data exchange systems [12] rely on dependencies to specify mappings. Given two schemas, $\mathbf{S}$ and $\mathbf{T}$, a *source-to-target tuple–generating dependency* (also called a s-t tgd or, equivalently, a tgd) is a first-order formula of the form $\forall \bar{x}(\phi(\bar{x}) \rightarrow \exists \bar{y}(\psi(\bar{x}, \bar{y}))$, where $\bar{x}$ and $\bar{y}$ are vectors of variables, $\bar{x}$ are universally quantified variables and $\bar{y}$ are existentially quantified variables. The body ϕ is a conjunctive query (CQ) over $\mathbf{S}$ and the head ψ is a CQ over $\mathbf{T}$.

Example 1. Consider Fig. 1 in which two schemas describing two scientists' local data are depicted. A set of correspondences v_1, v_2 and v_3 connects elements in the two schemas.

We report below two examples of s-t tgds for the two schemas above:

SOURCE-TO-TARGET TGDS

m_1. $\forall n, l$: $Hospital(n, l) \rightarrow \exists I$: $HealthCareInst(n, I)$

m_2. $\forall n, s, a, pi, l$: $Doctor(n, s) \land Grant(a, pi, n) \land Hospital(pi, l)$
  $\rightarrow \exists I$: $HealthCareInst(pi, I) \land Grant(a, n, I)$

A *schema mapping* is a triple $\mathcal{M} = (\mathbf{S}, \mathbf{T}, \mu_{\mathbf{st}})$ ($\mathcal{M}_{st}$, in short), where $\mathbf{S}$ is a source schema, $\mathbf{T}$ is a target schema, μ_{st} is a set of source-to-target tgds. If I is an instance of $\mathbf{S}$ and J is an instance of $\mathbf{T}$, then the pair $\langle I, J \rangle$ is an instance of $\langle \mathbf{S}, \mathbf{T} \rangle$. A target instance J is a *solution* of $\mathcal{M}$ and a source instance I (denoted $J \in \mathsf{Sol}(\mathcal{M}, I)$) iff $\langle I, J \rangle \models \mu_{st}$, i.e., I and J together satisfy the dependencies.

We distinguish between specific forms of s-t tgds, which are GAV (global-as-view) and LAV (local-as-view). A GAV tgd is a formula $\forall \bar{x}(\phi(\bar{x}) \rightarrow \mathcal{A}(\bar{x}))$, where the head is a single atom $\mathcal{A}(\bar{x})$. Similarly, a LAV tgd is a formula $\forall \bar{x}(\mathcal{A}(\bar{x}) \rightarrow \exists \bar{y}(\psi(\bar{x}, \bar{y}))$, where the body is a single atom $\mathcal{A}(\bar{x})$. GAV tgds are special cases of more general tgds, called GLAV, that contains conjunctions of atoms and existential variables in the head. Here and henceforth, we focus on GLAV mappings, thus expressed by means of GLAV s-t tgds in μ_{st}, as the others represent more restrictive cases. We denote such GLAV mappings with $\mathcal{M}$.

Example 2 (cont'd). Continuing with the above example, we define a schema mapping by the triple $\mathcal{M} = (\mathbf{S}, \mathbf{T}, \mu_{\mathbf{st}})$, where $\mathbf{S}$ is the source schema, $\mathbf{T}$ is the target schema and $\mu_{st} = \{m_1, m_2\}$. Moreover, $\mathcal{M}$ is a GLAV mapping.

We assume as customary that mappings among schemas are either provided by the users or by using external schema mapping tools.

We build on prior work [6,29] to define the semantics of query translation. Precisely, we denote with $inst(\mathbf{S})$ ($inst(\mathbf{T})$) the set of instances I (instances J, respectively).

Definition 1 (Semantics of query translation). *Suppose Q_i is a query posed against $\mathbf{S}$, and Q_j is a query posed against $\mathbf{T}$, $j \neq i$. Let Q_i^t denote a translation of Q_i against $\mathbf{T}$ and Q_j^t denote a translation of Q_j against $\mathbf{S}$. Then, Q_j^t is correct provided $\forall D_s \in inst(\mathbf{S})$: $Q_j^t(D_s) = Q_j(\mathcal{M}(D_s))$. The translation Q_i^t is correct provided $\forall D_t \in inst(\mathbf{T})$: $Q_i^t(D_t) = \bigcap_{D_s^k : \mathcal{M}(D_s^k) = D_t} Q_i(D_s^k)$.*

In other words, the translation Q_j^t is correct provided Q_j applied to the transformed instance $\mathcal{M}(D_s)$ and Q_j^t applied to D_s both yield the same results, for all $D_s \in inst(\mathbf{S})$. Notice that there may exist multiple solutions (there are so many solutions as there are instances D_s), but the semantics of certan answers adopt the intersection of all these solutions.

Also note that in this case, the direction of translation is *against* that of the mapping $\mathcal{M}$. As in [6], we henceforth call it *backward query translation*. This direction of translation is similar to view expansion, with $\mathcal{M}$ being the view definition. Translating a query Q_i posed against $\mathbf{S}$ to the schema $\mathbf{T}$ of peer p_j is *aligned* with the direction of the mapping $\mathcal{M}$, and represents the *forward query translation* [6]. The forward direction is more tricky, as the mapping $\mathcal{M}$ may not be invertible. In fact, there are two alternative strategies to make sense of this direction of translation: (i) obtaining the reverse schema mapping $\mathcal{M}^{-1}$ [3,11], such that the query rewriting semantics is the same as the backward direction; this strategy applies to the case in which one would like to recover the exchanged data, i.e. to find the source instance I from which the target instance J has been derived; (ii) focusing on the computation of a rewriting of a conjunctive query Q_1 over the source schema, assuming that a source instance I (D_s) is already available and adopting the semantics based on certain answers of all possible pre-images D_s^k; in such a case, it is possible to reuse the work done in the area of query answering using views for data integration [23,24].

Moreover, observe that in our setting we focus on query answering rather than on data exchange and on materializing a target instance. In fact, by following the semantics given in [6,13], we adopt the second strategy (ii), that lets us translate the query rather than the data and lets us realize query rewriting along the mappings[4]. This strategy is more natural in a P2P setting in which we do not need to reverse the mappings, and lets us avoid bringing the exchanged data back to the peers.

2.2   Network Model

We assume a heterogeneous network of peers $p_1, \ldots, p_n$, each peer having a distinct relational schema $S_1, \ldots, S_n$. Let $\mathcal{M}_{ij}$ be a generic GLAV mapping between a pair of schemas S_i, S_j, from peer p_i to peer p_j[5]. We assume that each peer has only one local schema, along with data defined according to the schema itself. Furthermore, for simplicity we ignore schema constraints in the query translation process.

We do not assume a symmetric distribution of the mappings, i.e. with a mapping $\mathcal{M}_{ij}$, we expect that either p_i (resp. p_j) stores the mapping or both of them. We have designed ad-hoc data structures to store mappings on each peer. Details on such data structures will be provided in Sect. 3.2.

2.3   Problem Statement

Given a mapping $\mathcal{M}$ from a peer p_i to a peer p_j, which we denote with $\mathcal{M}_{ij}$, $\mathcal{M}_{ij}$ is also called an *outward mapping* for p_i. The peer p_j is also called the *target peer* for this mapping. By opposite, a mapping $\mathcal{M}_{ji}$ from peer p_j to p_i is called an *inward mapping* for p_i (*outward* for p_j, resp.). Similarly, p_i is the target peer in such a case.

---

[4] Notice that we do not tackle the problem of merging results after applying the query rewriting to the acquaintance's database, but we take the union of these results.

[5] Notice that we do not also assume the existence of the mapping $\mathcal{M}_{ji}$, from peer p_j to peer p_i.

Let Q_i be an input query formulated at a peer in the network against an arbitrary schema S_i, and a direct outward mapping $\mathcal{M}_{ij} = S_i \rightarrow S_j$ (from S_i to S_j) and a direct inward mapping $\mathcal{M}_{ki} = S_k \rightarrow S_i$ (from S_k to S_i) and, in addition, transitively from S_j (resp. S_k) to any other reachable schema S_l (resp. S_m) for which it exists, without loss of generality, at least an outward mapping $\mathcal{M}_{jl}$ (resp. an inward mapping $\mathcal{M}_{mk}$) and so on, continuing from S_l and S_m to any other reachable schema through inward or outward mappings, then, the problem can be stated as:

– finding the relevant rewritings of Q_i along and against the direction of the mappings $\mathcal{M}_{ij}$ ($\mathcal{M}_{ki}$, resp.) and $\mathcal{M}_{jl}$ ($\mathcal{M}_{mk}$, resp.) and so on, by following the mappings which connect the schemas. All the relevant rewritings have to be computed by avoiding useless mapping paths from S_i to S_j, from S_k to S_i, from S_j to S_l, from S_m to S_k and so on, from any reached schema to any reachable schema connected by mappings.

Notice that the propagation of the input query Q_i to all peers in the network leads to collect as many rewritings as possible for that query. In fact, the input query Q_i can be certainly evaluated on the originating peer that hosts the schema S_i (upon which the query itself has been formulated) but may not be pertinent for all the schemas of other peers, unless relevant rewritings can be located. Moreover, the chosen strategy by which the results of the rewritten queries are conveyed towards the originating peer is a simple one, i.e. the results are unioned and possible duplicates are discarded. Alternatively, mapping composition could have been used here, but it falls beyond the scope of this paper.

The rewritings of Q_i follow the semantics given in Definition 1, whose correctness is proven in [6]. In this paper, we propose a query rewriting strategy different from the ones used in previous work [6,18] in which all possible translations are pursued, since we only exploit relevant translations. To this purpose, Sect. 3 introduces the notion of relevance of a query with respect to a mapping, and that of a relevant rewriting.

3 A Framework for Query Reformulation

In this section, we develop a novel framework for query reformulation in PDMS based on the notizion of social relevance. This framework relies on several contributions: a precise definition of relevance of a query wrt. a mapping; a new metric (AF-IMF) for computing such relevance and its supporting data structures; and a distributed method for computing AF-IMF in a P2P network.

3.1 Relevance of a Query wrt. a Mapping

To define such relevance, we consider a schema mapping scenario $\mathcal{M} = (\mathbf{S}, \mathbf{T}, \mu_{st})$, where $\mathbf{S}$ is a source schema, $\mathbf{T}$ is a target schema, μ_{st} is a set of source-to-target tgds that express the GLAV mapping.

Notice that here and henceforth we use *mapping rule* and *s-t tgd* as synonyms. Let $m \in \mu_{st}$ be a s-t tgd of the form $\forall \bar{x}(\phi(\bar{x}) \rightarrow \exists \bar{y}(\psi(\bar{x}, \bar{y})))$, with ϕ and ψ as CQ queries, containing the atoms $a_1(X_1), \cdots, a_n(X_n)$, with each X_i being an ordered set of parameters $(x_{i_1}, x_{i_2}, \cdots, x_{i_m})$, and each parameter being a variable $\$v_{i_k}$ (k=1,2,...,m) where $\$v_{i_1}$ is the variable for the paramenter x_{i_1} and so on.

Let Q be a CQ containing the atoms $a_1(X_1), \cdots, a_n(X_n)$, with each X_i being an ordered set of parameters $(x_{i_1}, x_{i_2}, \cdots, x_{i_m})$, and each parameter being a constant value c_{i_k} or a variable $\$v_{i_k}$ (k=1,2,...,m) where c_{i_1} is the constant value for the paramenter x_{i_1} or where $\$v_{i_1}$ is the variable for the paramenter x_{i_1} and so on.

A query atom $a_i(X_i)$ in ϕ (ψ, resp.) is unifiable with a query atom $a_j(X_j)$ in Q if a unifying substitution of variables and constant symbols exists. More precisely, a unification occurs if:

- (i) label(a_i) = label(a_j), i.e. both atoms have the same name;
- (ii) $\forall x_{i_k} \in X_i$, $\$v_{i_k}$ matches the variable symbol $\$v_{j_k} \wedge (i = j)$ or $\$v_{i_k}$ matches the constant symbol $c_{j_k} \wedge (i = j)$;
- (iii) variable bindings are consistent, i.e. $\forall x_{i_k} \in X_i$, if v_{i_k} matches a variable symbol v_{j_k}, it cannot match other variables or constants.

In other words, each query atom $a_i(X_i)$ must match an atom in the body (head, resp.) of a tgd with both its label and its set ordered of parameters ($x_{i_1}, \cdots, x_{i_m}$). Such a match follows the rules for atoms unification in Datalog (i.e. constant and variable unification).

Example 3. Consider again Fig. 1 and the mapping rules m_1 and m_2 specified in Sect. 2.

If we consider as a query $Q = Hospital(\$v_1, 'SanFrancisco')$, this query being posed against the source schema of Fig. 1, returns the names of all hospitals in San Francisco. Q consists of only one atom (*Hospital*) which has two parameters, a variable (i.e. $\$v_1$) plus a constant value (i.e. $'SanFrancisco'$).

We can now define the relevance of a query with respect to a mapping rule, as follows.

Definition 2 (Relevance Forward). *Given a schema mapping $\mathcal{M}_{ij}$ that maps elements of the schema S_i into elements of S_j and let m be a mapping rule in μ_{ij}, a conjunctive query Q posed against S_i along the direction of the mapping rule is relevant to m iff each atom appearing in Q can be unified with an atom in the body of m.*

Definition 3 (Relevance Backward). *Given a schema mapping $\mathcal{M}_{ij}$ that maps elements of the schema S_i into elements of S_j and let m be a mapping rule in μ_{ij}, a conjunctive query Q posed against S_j against the direction of the mapping rule is relevant to m iff each atom appearing in Q can be unified with an atom which only contains universally quantified variables in the head of m.*

Consequently, we can now define the relevance of a query wrt. the whole mapping, as follows.

Definition 4 (Mapping Relevance). *Let* $\mathcal{M} = (\mathbf{S}, \mathbf{T},\ \mu_{st})$ *be a mapping, where* $\mathbf{S}$ *is a source schema,* $\mathbf{T}$ *is a target schema and* μ_{st} *be the set of s-t tgds and let* $m \in \mu_{st}$ *be a mapping rule of such mapping. A query* Q *posed against the mapping* $\mathcal{M}$ *is relevant if there exists at least one mapping rule* $m \in \mu_{st}$ *so that* m *is forward or backward relevant for* Q.

Example 4 (cont'd). Continuing with the example above, shown in Fig. 1, it is easy to check that the query Q above is forward relevant to both m_1 and m_2, according to the above definition. If we consider a query $Q' = Grant(\$x, \$y, \$z)$ and a query $Q'' = HealthCareInst(\$y, \$z)$, neither Q' nor Q'' are backward relevant to either mapping rule.

Here and henceforth, we will use the term *relevance* to denote mapping relevance, unless otherwise specified. It follows that, if a query Q is relevant to a mapping $\mathcal{M}_{ij}$, its translation Q_i^t is also relevant to that mapping.

Proposition 1. *If a query* Q *formulated against* S_i *is relevant to a mapping* $\mathcal{M}_{ij}$, *its translation* Q^t *formulated against* S_j *is also relevant to* $\mathcal{M}_{ij}$, *and viceversa.*

Proof. To prove the above proposition, we have to consider both directions of translations, i.e. forward and backward and apply the corresponding definition of relevance. We start with the forward direction of translation and the forward relevance. Being the mapping μ_{ij} a tgd of the form $\forall \overline{x}(\phi(\overline{x}) \rightarrow \exists \overline{y}(\psi(\overline{x}, \overline{y}))$, by applying Definition 2, the query Q must contain all constants and variables, that by substitution can be matched to at least one variable in the vector of variables $\overline{x}$. It follows that the query translation from Q to Q^t brings at least one variable to the head $(\psi(\overline{x}, \overline{y}))$, thus yielding at least an atom in Q^t, that contains that variable. This implies that Q^t exists (as the set of certain answers to the query Q is not empty) and is also relevant to the mapping μ_{ij} since it contains at least one atom in ψ. We now show that the same holds for the backward direction of translation and backward relevance. In such a case, we assume that the query Q is posed against the source schema S_j, thus, being the mapping μ_{ij}, by applying Definition 3, the query must contain all constants and variables that by substitution can be matched to at least one variable v in the vector of variables $\overline{x}, \overline{y}$. There are two possible cases: (1) $v \in \overline{x}$; (2) $v \in \overline{y}$. Case (1) implies that Q^t exists and the view expansion is not empty. Henceforth, Q^t is also relevant to the mapping μ_{ij} since it contains at least one atom in ϕ. Case (2) leads to an empty query translation, thus $Q^t = \emptyset$, which is trivially not contained in ϕ, as it contradicts the definition of backward relevance. $\square$

The above query Q entails a *relevant rewriting*, according to the next definition.

Definition 5 (Relevant Rewriting of a Query). *Given a query* Q *relevant to a mapping* $\mathcal{M}_{ij} = S_i \rightarrow S_j$, *its translation* Q^t *is a relevant rewriting of* Q *against* S_j. *We say that* $\mathcal{M}_{ij}$ *rewrites* Q *into* Q^t, *denoted by* $Q \overset{\mathcal{M}_{ij}}{\rightarrow} Q^t$.

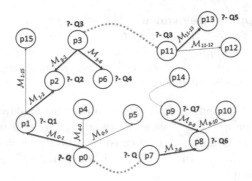

Fig. 2. (a) Useless rewriting sequence, (b) alternative rewriting sequences and relevant mappings (in bold).

Based on the above definition, we can now define a *rewriting sequence*, as follows.

Definition 6 (Rewriting Sequence). *For a query Q, if $Q_0 \overset{\mathcal{M}_{01}}{\rightsquigarrow} Q_0{}^t = Q_1 \overset{\mathcal{M}_{12}}{\rightsquigarrow} Q_1{}^t = Q_2 \cdots \overset{\mathcal{M}_{(n-1)n}}{\rightsquigarrow} Q_{n-1}{}^t = Q_n$, we say that Q rewrites into $Q_n{}^t$. The mappings $\mathcal{M}_{01}, \cdots, \mathcal{M}_{(n-1)n}$ are called the rewriting sequence.*

An example of *rewriting sequence* starting from the peer p_0 to the peer p_7 is highlighted in bold in Fig. 2 (b).

According to our problem definition, we need to find all the possible rewriting sequences of a given input query Q_0 on the initiating peer p_0. However, a rewriting sequence might not always exist between p_0 and an arbitrary peer p_n, since there might be an intermediate mapping that does not entail a relevant rewriting of the query. We denote such mapping as a *useless mapping* and the entire sequence a *useless rewriting sequence*. An example of such a sequence is depicted in Figure 2 (a), from p_0 to p_3, where the mapping from peer p_1 to p_2 is not relevant. Avoiding useless sequences is quite straightforward because they can be detected by adopting a local metric to assess whether the target of the current peer is able to handle the query, before actually shipping the query itself to that target. Such evaluation can be done by using the mapping rules themselves, as they are locally stored on the current peer and can be easily inquired to that purpose.

Another issue that often occurs is that of alternative rewriting sequences, as depicted in Fig. 2 (b). Indeed, the current peer may have multiple alternative paths to rewrite a given query, and may have to choose the most appropriate one. E.g. in Fig. 2 (b), p_0 could choose among three possible alternatives p_1, p_4 and p_5. Exhaustively pursuing all possible rewritings is obviously not feasible, due to the great number of destination peers and rewriting sequences. Moreover, only fews rewritings along the sequences may happen to be the most relevant ones, which is always preferable to pursue. To this purpose, we need a *relevance score* for each possible rewriting sequence (described next) in order to be able to rank the possible rewriting alternatives. Consequently, it becomes feasible to rewrite the queries along the most relevant paths (e.g. represented by the bold arrows in Fig. 2 (b)).

Remark. We observe that one could apply Definition 4 in a straightfoward manner to address the previous problems. However, a relevance score solely based on a local metric would not be sufficient as it would only check one mapping at a time. Conversely, one needs to check an entire rewriting sequence among the possible alternatives. Thus, a global metric is needed to assess the relevance of queries with respect to the mappings in a rewriting sequence.

Moreover, the above definitions only handle the cases of mappings that contain *all* the atoms of a query Q in their body or in their head, whereas the total number of remaining non-relevant atoms in the mapping body or head is also important, as we show in the following example.

Example 5 (cont'd). If we consider the query Q of Example 3, the mappings m_1 and m_2 of Example 1 present a different degree of relevance with respect to Q. In particular, m_1 is more relevant to the query than m_2 itself, as the query atoms exactly coincide with the atoms in the body of the mapping m_1.

To make the above example more general, let us assume that for a given query Q triggered on a peer there are n distinct mappings $(m_1, \cdots, m_n)$ with as many distinct peer friends $(p_1, \cdots, p_n)$. In order to avoid useless query translations, the triggering peer has to choose which among $m_1 \cdots m_n$ is (are) the best mapping(s) to exploit for query translation. This simple example already motivates the importance of establishing a degree of relevance for a set of candidate mappings with respect to a given query.

3.2   Relevance Metric

In this section, we present our novel relevance metric to quantify the degree of relevance of mappings in the network and the data structures that allow computing it. We highlight that the metric we are introducing is able to take into account both the specificity of a mapping rule w.r.t. a query Q and the overall importance of the query atoms in the network.

AF-IMF Metric. Our metric which we call AF-IMF, i.e. *atom frequency, inverse mapping frequence,* is an adaptation of the classical information retrieval metric TF-IDF to schema mapping. Variations of the TF-IDF weighting scheme are often used by search engines as a central tool for scoring and ranking a document's relevance given a user query. Similarly, AF-IMF is a statistical measure to evaluate how important a query atom is to a mapping in the entire collection. The importance increases proportionally to the number of times an atom appears in the mapping but is offset by the frequency of the atom in the collection.

In the following, we first define the AF-IMF for an individual mapping rule, then we extend it to entire mappings.

We introduce the *atom count* in the given mapping rule m, as the number of times a given query atom a_q fully appears in m by using constant and variable unifications. This count is usually normalized by the number of occurrences of all atoms in m. We assume that each atom can only appear once in a mapping rule, thus implying that the atom frequency can be approximated to 1.

$$\text{AF}_{i,j} = \frac{n_{i,j}}{\sum_k n_{k,j}} \simeq \frac{1}{k}$$

where $n_{i,j}$ is the number of occurrences of the considered atom a_i in m_j, and the denominator is the sum of number of occurrences of all k atoms occurring in the body (head, resp.) of m_j, where a_i respectively appears. Notice that having two separate AF on the body and head according to where the atom a_i appears in the mapping rule m_j is crucial to characterize the forward from backward relevance, respectively.

The inverse mapping frequency is a measure of the general importance of the atom, obtained by dividing the total number of mapping rules by the number of mapping rules containing the atom in the body (head, resp.), and then taking the logarithm of that quotient.

$$\text{IMF}_i = \log \frac{|M|}{|\{m_j : a_i \in m_j\}|}$$

with $|M| = |\cup_{i=1\ldots n}\,\mu_{st}|$ being the total number of mapping rules in the network, which amounts to the union (without duplicates) of all the source-to-target tgds; and $|\{m_j : a_i \in m_j\}|$ being the number of mapping rules where the atom a_i appears (that is $n_{i,j} \neq 0$) in the body (head, resp.). If the atom is not in the network, this will lead to a division-by-zero, thus it is common to use $1 + |\{m_j : a_i \in m_j\}|$ instead.

Notice that the computation of AF depends on both the current query atom a_i and the current mapping rule m_j. Differently, the IMF computation does not depend on the current mapping rule m_j but only on the current query atom a_i.

Then,

$$(\text{AF-IMF})_{i,j} = \text{AF}_{i,j} \times \text{IMF}_i \simeq \frac{1}{k} \times \text{IMF}_i$$

The above formula implies that the mapping rules with less atoms are preferred with respect to those with more atoms. Therefore, a high weight in AF-IMF is reached by mapping rules with low total number of atoms, and low frequency in the global collection of mapping rules.

What has been already observed above on forward from backward relevance implies that a different value of the AF-IMF is computed for atoms appearing in the body (head, resp.) of the mapping rules in a similar fashion.

A further step would lead to extend the above metric for the query atoms a_q altogether so that it is possible to assign a comprehensive value of relevance the entire query Q with respect to the mapping rule m_j (as opposed to the previous case, when only an individual query atom a_i was considered). Such step implies a simple measure (e.g. the *sum*) to put together the AF-IMF scores separately obtained by the query atoms a_q of Q.

After applying the composition of the above scores, we obtain the overall score for the mapping rule m_j, as in the following:

$$(\text{AF-IMF})_j = \sum_i (\text{AF-IMF})_{i,j}$$

After defining the notion of AF-IMF for an individual mapping rule m, we now extend the definition to the entire mapping $\mathcal{M}$. We recall that the final goal of our metric is to assign a relevance value to those mappings that the current peer is about to evaluate in order to realize the query translation of Q.

Being a mapping scenario $\mathcal{M} = (\mathbf{S}, \mathbf{T}, \mu_{st})$ defined by means of a set of K ($K > 0$) mapping rules in μ_{st}, we compute the overall AF-IMF score for $\mathcal{M}$ as the sum of the AF-IMF scores obtained by each mapping rule $m \in \mu_{st}$ (according to the forward or backward definition of relevance).

In other words, if the relevance is backward the query Q matches the head side of the mapping rule m_j (see Definition 3), the AF-IMF computation is done as shown below:

$$(\text{AF-IMF})_{\mathcal{M},\text{head}} = \sum_{j=1}^{K_h} (\text{AF-IMF})_j$$

where $K_h \subseteq K$ is the number of rules $m_j \in \mu_{st}$, such that Q matches their head side.

Instead, if the relevance is forward the query Q matches the body side of the mapping rule m_j (see Definition 2), the AF-IMF computation is done as shown below:

$$(\text{AF-IMF})_{\mathcal{M},\text{body}} = \sum_{j=1}^{K_b} (\text{AF-IMF})_j$$

where $K_b \subseteq K$ is the number of rules $m_j \in \mu_{st}$, such that Q matches their body side.

The overall relevance of the query Q with respect to the entire mapping $\mathcal{M}$ is the maximum value between the two formulas above:

$$(\text{AF-IMF})_{\mathcal{M}} = \max((\text{AF-IMF})_{\mathcal{M},\text{head}}, (\text{AF-IMF})_{\mathcal{M},\text{body}})$$

In such a way, given a query Q as input, the AF-IMF metric assigns a score of relevance to each inward and outward mapping of the peer, to let it choose the most relevant paths for query translation, i.e. the ones with the highest scores.

Example 6. Consider Fig. 3 that is a slightly different version of Fig. 1. A set of correspondences v_1, v_2, v_3, v_4 and v_5 connects elements in the two schemas. Assume that the set of corresponding s-t tgds is the one reported below:

SOURCE-TO-TARGET TGDS
m_1. $\forall n, l$: $Hospital(n, l) \rightarrow \exists I$: $HealthCareInst(n, I)$
m_2. $\forall n, s, d, dw, da, a, pi, l$: $Doctor(n, s, d) \wedge Grant(a, pi, n)$
     $\wedge Department(d, dw, da) \wedge Hospital(pi, l)$
     $\rightarrow \exists I, g$: $HealthCareInst(pi, I) \wedge Grant(a, n, I, d)$
     $\wedge Dept(d, g, dw)$
m_3. $\forall n, w, a$: $Department(n, w, a) \rightarrow \exists g$: $Dept(n, g, w)$

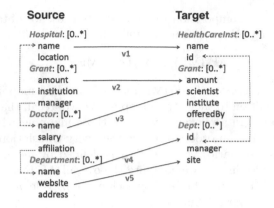

Fig. 3. A Schema Mapping Example

The mapping $\mathcal{M}$ among source S and target T includes all the above three mapping rules.

Now, let us imagine that the source peer S is connected to other target peers (T_1, T_2 and T_3) all having, for simplicity, an identical target schema T with sets of different mappings. Such mappings ($\mathcal{M}_1$, $\mathcal{M}_2$ and $\mathcal{M}_3$) are simply variants of $\mathcal{M}$, i.e. mappings derived from $\mathcal{M}$ by including a different subset of the mapping rules of μ_{st}, as specified in the following:

- $\mathcal{M}_1 : \mu_{st} = \{m_1, m_2\}$
- $\mathcal{M}_2 : \mu_{st} = \{m_1, m_3\}$
- $\mathcal{M}_3 : \mu_{st} = \{m_2, m_3\}$

If we compute the AF-IMF scores for all the mappings above, i.e. $\mathcal{M}$, $\mathcal{M}_1$, $\mathcal{M}_2$ and $\mathcal{M}_3$, it is easy to check that $\mathcal{M}$ will always get the highest score, since it is the most complete mapping. Therefore the peer T, that is connected to S through $\mathcal{M}$, represents the most relevant peer to follow in the query reformulation process.

Nevertheless, a further complication arises since IMF cannot be exactly computed as the size of the entire collection of mapping rules at a given time is not known, due to the fact that the network is dynamically changing.

To address this problem, each peer is equipped with a set of semantic data structures, that summarizes the local and external mappings of a peer (see next Section for details). Thus, by exploiting such data structures, we can compute an approximation of IMF for the distributed case, as discussed in Sect. 3.2.

Semantic Data Structures. In this section, we first introduce the local semantic data structures stored on each peer. Then, in Sect. 3.2, we present how they can be exploited to approximate the IMF values.

Local Data Structures on a peer

Local Semantic View (LSV)

Mapping-content

Atom	Mapping	SrcP	TgtP	Peer
Hospital	SHA−1(m1)	p1	p3	p5
Hospital	SHA−1(m2)	p3	p4	p6
Hospital	SHA−1(m2)	p3	p4	p6
Grant	SHA−1(m2)	p3	p4	p6
....	....	....	....	....

View

Peer	Age
p1	40
p3	0
p4	1

Mapping Summary

Atom	Count(m)
Hospital	2
Grant	1
.....	....

Fig. 4. Local data structures on a peer.

Figure 4 represents the local data structures on each peer. Each peer maintains a set of local or internal mapping rules[6], i.e. mapping rules from its local schema to the schema of each of its acquaintances, the latter being a selected subset of the peer's neighbors [16,20]. Moreover, it also stores a *Local Semantic View* (LSV in short), that encloses information about external mapping rules (distinct from the local ones), selected uniformly at random from the network. This view is used to compute the relevance values. Precisely, an LSV for each peer consists of: a five-column table *Mapping-content (Atom, Mapping, SrcPeer, TgtPeer, Peer)*, and of a two-column table *View (Peer, Age)*, with a foreign key constraint between *View.Peer* and *Mapping-content.Peer*. The *Mapping-content* relation has a column *Atom* containing the atom of a mapping rule in the network; a column *Mapping* containing the ID of the mapping rule in which *Atom* appears; a column *SrcPeer* containing the ID of the external source peer to which *Mapping* is an outward mapping; a column *TgtPeer* containing the ID of the external target peer to which *Mapping* is an inward mapping; a column *Peer* containing the ID of the peer in the network that has provided the current tuple in a gossip cycle. The *View* relation has a column *Peer* containing the ID of a peer in the network; a column *Age* containing a numeric field that denotes the age of the mapping rules since the time in which they have been included within the *View*. Figure 4 shows an example of a LSV on a peer.

To uniquely identify each mapping rule in the PDMS, we assign an ID to each mapping, using cryptographic hash functions (e.g. SHA-1) to reduce the probability of collision[7].

---

[6] The mapping statements have been omitted from Fig. 4 to avoid clutter.

[7] In DHTs or structured P2P networks, on which PDMS are based, a unique key identifier is assigned to each peer and object. IDs associated with objects are mapped through the DHT protocol to the peer responsible for that object. In our setting, each object is a mapping.

As the size of the LSV is limited, it implies that the view entries need to be replaced, based on their age information. In order to maintain each LSV on the peers, we adopt classical thread-based gossiping mechanisms, aiming at updating the LSV with newly incoming tuples from the outside. In Sect. 4.4 we provide the details of such maintenance.

Besides local mappings, each peer also maintains an additional descriptive data structure of such mappings, called *Mapping Summary*, which is implemented as a local Bloom filter [5]. A Bloom filter is a method for representing a set $A = \{a_1, a_2, \cdots, a_n\}$ of n elements (also called keys) to check the membership of any element in A. In the *Mapping Summary*, a bit vector v of m bits, initially set to 0, represents the positions of k independent hash functions, $h_1, h_2, \cdots, h_k$, each with range $\{1, \cdots, m\}$. In the *Mapping Summary*, a key is built as follows: for each mapping rule, the conjunction of all atoms in the body ϕ (in the head ψ, resp.) is a key; each individual atom a_i in the body ϕ (in the head ψ, resp.) is a key; each subset $a_1, \ldots, a_n$ of the atoms in the body ϕ (in the head ψ, resp.), such that it exists at least one joined variable in each atom a_i and a_{i+1}, is a key. By enumerating the above keys, each body (head, resp.) of the mapping rule has a total of $\frac{n(n+1)}{2}$ entries in a *Mapping Summary*. Such a combination of atoms and/or individual atoms may appear in several distinct local mapping rules on that peer. For each atom (or combination of atoms thereof) $a \in A$, the bits at positions $h_1(a), h_2(a), \cdots, h_k(a)$ in v are set to 1. A membership query checks the bits at the positions $h_1(a), h_2(a), \cdots, h_k(a)$. If any of them is 0, the atom a is not in the set. Otherwise, we conjecture that a is in the set, although this may lead to a false positive. The aim is to tune k and m so as to have an acceptable probability of false positives. The advantage of using Bloom filters resides in the fact that they require very little storage, at the slight risk of false positives. Such probability is quite small already for a total of 4 different hash functions [15]. Figure 4 shows an example of a *Mapping Summary* on a peer.

Similarly to the LSV, the *Mapping Summary* needs to be maintained in the presence of changes of the atoms within the mapping rules, and/or additions and deletions of the mapping rules themselves. This is done by maintaining in each location l in the bit vector v, a count $c(l)$ of the number of times that the bit is set to 1. The counts are initially all set to 0. When insertions or deletions take place, the counts are incremented or decremented accordingly.

Finally, to allow friendship linking among peers, a peer mantains a third structure, that is basically a local *FOAF file* containing the URIs of its friends FOAF files. Whenever a user (or a peer) generates its FOAF file, it can obtain an identity for that file in the form of a URI. This URI could point to a reference in a friend's FOAF file. URIs correspond to unique peer and object identifiers in a PDMS. In particular, a peer p_1 may need to store into its FOAF file: (1) the list of other peers he knows and he is friend with, as a link to its friend's FOAF file (e.g. P3.rdf in the example); (2) possibly, the link to its friend's *Mapping Summary (e.g. P3_MapSum in the example below)*.

```
<foaf:knows>
  <foaf:Peer>
    <foaf:peerID> P3</foaf:peerID>
        <rdfs: seeAlso rdf: resource =
            'http://www.mirospthree.com/P3.rdf'/>
        <rdfs: seeAlso rdf: resource =
            'http://www.mirospthree.com/P3_MapSum'/>
  </foaf:Peer>
</foaf:knows>
```

The main goal of FOAF files is to maintain the current friendship links of a given peer. During query translation, the FOAF file is expanded by adding new friends, by invoking the Algorithm FindDirectFOAFFriends, described in Sect. 4. Notice that adopting and exploiting the friendship links of a given peer during the query translation process is complementary to exploiting the semantic mappings towards the peer's acquaintances. In fact, the friendship links are especially useful in the presence of network churn, as they act as a background network regardless of the peer's acquaintances and its direct inward/outward mappings. A more detailed experiment about network churn, scalability and the usefulness of FOAF links is provided in Sect. 5.

In our model, no peer can access the other peer's mapping summary until an explicit friendship link has been established between such peers, thus leading to modify their respective FOAF files accordingly. This mechanism gracefully replaces an explicit negotiation and coordination among peers for accessing their respective data structures. An additional access control mechanism, e.g. [8], can be adopted on top of FOAF files to further strenghten the security of the network. More sophisticated privacy and security mechanisms are beyond the scope of our paper.

In the remainder of this discussion and in Sect. 4, we denote the peers indexed in a FOAF file as 'friends'. These represent the peers whose mapping summary can be accessed, in order to widen the scope of the queries. In particular, in Sect. 4, we will discuss how to enlarge the set of simple friends of a peer by exploiting friendship links in its FOAF file.

Distributed Computation of AF-IMF. Using the local semantic view and the local mapping rules, we can compute IMF_i distributively, as follows. Let k be the number of distinct local mapping rules entries and let t the number of distinct mapping rules in the LSV. We know by definition that the k entries and t entries are not overlapping, thus we may say that locally we have $k + t$ mapping rules. Then, we have to determine what is the approximation of $|M|$, the total number of mapping rules in the collection, possibly without duplicates. We may think of computing N, the total number of peers in the network and multiplying it by $k + t$, thus obtaining $|M| = (k + t) \times N$. Moreover, we observe that N can be easily computed if we know the network topology. For instance, for DHTs it suffices to record the size of the routing table, which is $r = log(N)$, and by taking the inverse as $2^r = N$. For super-peer networks, we may have an entry point that registers the total number of peers N. For unstructured P2P networks, we may rely on flooding to count the total number of peers in the network. In a similar way, the $|\{m_{ij} : a_i \in m_{ij}\}|$ can be computed by selecting among the k and t local mappings, those that contain the atom a_i, thus obtaining $|\{m_{ij} : a_i \in m_{ij}\}| = (k_i + t_i) \times N$.

However, we need to avoid duplicate mapping rules in the previous computation. In order to do this, we need to uniquely identify a mapping in the entire network. A simple and effective way to do this is to couple each mapping with its signature, using a cryptographic hash function (e.g. SHA-1). We present in Sect. 4 an algorithm to compute AF-IMF distributively, that avoids duplicate mappings by using signatures.

Remark. As a final observation, we underline that the problems illustrated in Fig. 2 are both overcome, since the useless sequences do not affect the AF-IMF metric. Moreover, AF-IMF enables the search of the most relevant rewriting sequences in a global fashion, as expected by our previous reasoning. In the experimental analysis (Sect. 5), we show the effectiveness of this metric, also when compared to a local metric (e.g. by adopting the sole AF as a local metric).

4   Algorithms

In this section, we illustrate our algorithm: the core algorithm that translates a query based on relevance; an algorithm for seeking new friends that contain relevant mappings for the query; a distributed algorithm to compute the relevance of mappings, that is used by the two former algorithms. Finally, we briefly discuss the gossiping algorithm for updating semantic views.

4.1   Distributed Computation of the Relevance

Algorithm 1 computes a measure of the relevance of a set of mapping rules on a given peer with respect to an input query, with the aim of getting an ordered top-k list of mappings to be exploited (by Algorithm 2) and the aim of finding new friends by (Algorithm 3).

The algorithm has two main parts. Lines 1-14 aim at computing the IMF values for each query atom, and this entails a separate computation, depending on which side of the mapping rule the query atom belongs. Therefore, two vectors $BodyIMF$ and $HeadIMF$ are built to store the IMF values of each atom in the query Q.

Then, the second part of the Algorithm (lines 15-32) computes the denominator of AF values as the total number of atoms appearing in the matched side of the mapping rule, and the complete value of AF-IMF is then returned. The final relevance value (line 32) for the whole mapping rule is taken by applying a suitable ranking function to the values in the above vectors (e.g. *sum*).

Let us observe that the computation of the IMF only depends on the atom a_i in the query Q, and not on the current mapping rule. For this reason, we also make sure that the computation at lines 1-14 is done only once for the same query, by saving intermediate results.

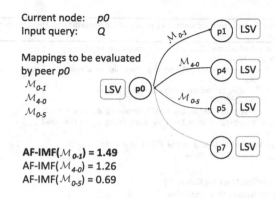

Fig. 5. An example of ComputeRelevance (Algoritm 1).

Indeed, the computation is done by asking each known peer (both destination peers through mappings and new discovered peer friends in the FOAF file f). The maximum number of inquiries is given by the $REQS$ threshold. Observe that if $REQS = 0$ no external inquiries have been done, and only the entries of the current peer's LSV have been inspected, whereas a value of $REQS$ greater than 0 leads to also inspect the LSV of external peers. Also note that such inquiries are done by discarding duplicates through the asynchronous method GetDistinctMappingRules, that checks the signatures of the mapping rules. We omit the pseudo code of this method for space reasons.

Figure 5 shows an example of how Algorithm 1 computes the relevance. A query Q is initially posed against the peer p_0, which in turn chooses among three alternative target peers (also called acquaintances). Also, note that from p_0 toward p_7 there is no direct mapping, but rather a FOAF link depicted by a dotted blue arrow. Thus, mappings $\mathcal{M}_{01}$ (from p_0 to p_1), $\mathcal{M}_{40}$ (from p_4 to p_0) and $\mathcal{M}_{05}$ (from p_0 to p_5) must be evaluated aiming at finding the *top-k* relevant ones for the input query (in this example, we assume for simplicity that $k = 1$). By inspecting p_0's LSV, Algorithm 1 performs the computation of the relevance metric for each mapping rule m of each mapping involved ($\mathcal{M}_{01}, \mathcal{M}_{40}, \mathcal{M}_{05}$), by assigning an AF-IMF value to each involved atom, as previously discussed. At the end, the mapping $\mathcal{M}_{01}$ (from p_0 to p_1) gets the highest relevance score amongst all the other mappings, thus becoming the *top-1* step in the rewriting sequence of query Q.

4.2   Translating Queries Based on Relevance

Algorithm 2 translates a query initiated at a peer, first against its set of local mappings and then by exploiting local friendship links at that peer.

Notice that the core of the algorithm (lines 10-30) is essentially the translation algorithm reported in [6] that we have extended and improved by considering our relevance metric and by exploiting FOAF friendships. Thus, a formal proof of correctness of Algorithm 2 directly follows from the proof of correctness of

---

Algorithm 1: ComputeRelevance - computes the relevance according to the gossiped information in the local semantic view

---

    Input : A query Q as set of atoms $\mathcal{A}_Q$, a list of k mapping rules m_k,
    a peer p with its LSV and FOAF file f
    Output: The vector of relevance values RV for the input list of k mapping rules

1 **foreach** *atom a_i in $\mathcal{A}_Q$* **do**
     //Compute the IMF value according to the *matchedSide*
2   n = total nr. of mapping rules in the LSV;
3   nB_i = total nr. of mapping rules in the LSV containing a_i in the body;
4   nH_i = total nr. of mapping rules in the LSV containing a_i in the head;
5   Let $count_{reqs}$ = 0;
6   **foreach** *p' in the View of LSV and in the FOAF file f* **do**
7   **if** $count_{reqs} >= REQS$ **then**
8   break;

9   n += **GetDistinctMappingRules**(p');
10   nB_i += **GetDistinctMappingRules**(p', "Body", a_i);
11   nH_i += **GetDistinctMappingRules**(p', "Head", a_i);
12       $count_{reqs}$++;

13     $BodyIMF[i]$ = log(n / (1 + nB_i));
14     $HeadIMF[i]$ = log(n / (1 + nH_i));

15 **foreach** *mapping rule m_k in the list of input mapping rules* **do**
16   **if** *all atoms a_i in $\mathcal{A}_Q$ are in the body of m_k* **then**
17      $matchedSide$ = "Body";
18   **else**
19   **if** *all atoms a_i in $\mathcal{A}_Q$ are in the head of μ_k* **then**
20        $matchedSide$ = "Head";
21   **else**
       //No relevance
22        $RV[k]$ = 0;
23   continue;

24    $BodyAF_i$ = total nr. of atoms in the body of m_k
25    $HeadAF_i$ = total nr. of atoms in the head of m_k
26   **foreach** *atom a_i in $\mathcal{A}_Q$* **do**
27      $AF - IMF[i]$ = 0;
      //Compute the AF-IMF value for a_i according to the *matchedSide*
28   **if** *matchedSide == "Body"* **then**
29        $AF - IMF[i]$ = (1 / $BodyAF_i$) * $BodyIMF[i]$;
30   **else**
31        $AF - IMF[i]$ = (1 / $HeadAF_i$) * $HeadIMF[i]$;

    //Compute the final relevance value RV for the whole mapping rule m_k
32    $RV[k]$ = RankFn($AF - IMF[i]$)

33 **return** RV;

---

the translation algorithm in [6] (cfr. Theorem 1) that we omit for the sake of conciseness.

The algorithm is inherently recursive, and at each iteration increases the number of *query hops*, until a given threshold α is reached. This avoids exploring the entire network, by conveying the query toward a limited number of peers. By exploiting the notion of relevance for the input query Q, new *friends* are discovered and added to the FOAF friend list.

By invoking the method FindDirectFOAFFriends (line 4), the current peer enlarges the list of its friends in its local FOAF file. Therefore, new relevant friends might be discovered, similarly to real-life friendship mechanisms, and to friend-bases game applications (e.g. Farmville) in modern social platforms (e.g. Facebook). Lines 6-8 invoke the method ComputeRelevance for each local

---

Algorithm 2: TranslateQuery - Query translation based on relevance

Input : Query Q as set of atoms $\mathcal{A}_Q$ and a peer p with its list $MList$ of local mappings $\Sigma_{1 \leq i \leq n} \mathcal{M}_i$, and its FOAF file f

Output: Query results res of the query Q against the peer p exploiting both the set of local relevant mappings $\Sigma_{1 \leq i \leq n} \mu_i$ and new relevant peer friends

  1 **if** $Q.query\text{-}hops \geq \alpha$ **then**
  2 | return res;

  3 increase $Q.query\text{-}hops$ by 1;
  4 Call **FindDirectFOAFFriends**(Q, p);
  5 Let L be a list of mappings ordered by relevance;
  6 **foreach** *local mapping* $\mathcal{M}_i$ *in* $MList$ **do**
  7 | RV = Call **ComputeRelevance**$(Q, \mathcal{M}_i.MappingRules())$;
  8 | $mapscore[i]$ = SumValuesFromVector(RV);

  9 L = **Order** $MList$ according to mapping relevance values in $mapscore$;
10 **foreach** *top-k ordered mapping* $\mathcal{M}_i$ *in* L **do**
11 | **if** $\mathcal{M}_i$ *has been already processed* **then**
12 | | continue;
       | //To avoid cycles

13 | Let $destPeer$ the destination peer through mapping $\mathcal{M}_i$;
14 | **if** Q *is relevant to the body of* $\mathcal{M}_i$ **then**
15 | | Translate Q along $\mathcal{M}_i$ obtaining Q'
16 | | **if** $\mathcal{M}_i$ *is outward* **then**
17 | | | $res = res \cup$ Eval(Q);
18 | | | $res = res \cup$ TranslateQuery$(Q', destPeer)$;
19 | | **else**
20 | | | $res = res \cup$ Eval(Q');
21 | | | $res = res \cup$ TranslateQuery$(Q, destPeer)$;

22 | **else**
23 | | **if** Q *is relevant to the head of* $\mathcal{M}_i$ **then**
24 | | | Translate Q against $\mathcal{M}_i$ obtaining Q'
25 | | | **if** $\mathcal{M}_i$ *is outward* **then**
26 | | | | $res = res \cup$ Eval(Q');
27 | | | | $res = res \cup$ TranslateQuery$(Q, destPeer)$;
28 | | | **else**
29 | | | | $res = res \cup$ Eval(Q);
30 | | | | $res = res \cup$ TranslateQuery$(Q', destPeer)$;

31 Let F be a list of friends;
32 F = Call **ComputeFriendsWithGreatestCount**(Q, f);
33 **foreach** *top-k ordered friend* $pFoaf$ *in* F **do**
       | //To exploit new interesting peer friends
34 | $res = res \cup$ TranslateQuery$(Q, pFoaf)$;

35 **return** res;

---

mapping $\mathcal{M}_i$ of the peer, in order to get the relevance scores for such local mappings. Then, at line 9, the list of local mappings is ordered according to the calculated relevance scores with respect to the input query Q.

Mapping identity is checked in lines 10-11, in order to avoid using the same mappings more than once in different iterations. The query rewriting proceeds by taking into account the direction of the mapping (cfr. Definition 1) and then can take place *along* (line 12) or *against* (line 21) the mapping, thus obtaining the translated query Q'. Then, according to the type of the mapping considered - if inward or outward, the input query Q and the translated one Q' are executed against the current peer or instead used in the recursive call of the Algorithm. Next, the query translation task is pushed towards the new

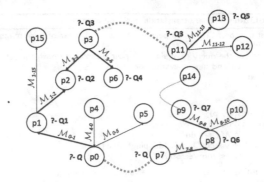

Fig. 6. An example of TranslateQuery (Algorithm 2).

interesting peer friends encoded in the FOAF file f (lines 31-34). This search exploits the peer friends' *Mapping Summary* to check whether there is a high number of mappings that contains atoms of the input query Q (via the method ComputeFriendsWithGreatestCount). Finally, all the query results *res* are returned (line 34) as the union of all the results harvested throughout the recursive invocations of the algorithm.

Figure 6 shows an example of execution of Algorithm 2. A query Q is posed against the peer p_0. In trying to choose the most relevant rewriting sequence (lines 5-30), p_0 applies Algorithm 1 (for simplicity, we assume that $top\text{-}k = 1$). This way, the query Q is rewritten and traslated transitively until p_6 is reached. No translation is further possible, since p_6 is a terminal node. However, FOAF links found in line 4 of the Algorithm 2 are also exploited in this example. Indeed, they allow to traverse disconnected subsets of the nodes in the graph of Fig. 6 (lines 31-34). If the friend reachable through the link is able to treat the query, the query can be further propagated to that friend and its subgraph. In the figure, one can see that p_7 and p_{11} receive the queries Q and $Q3$ respectively from p_0 and p_3. By contrary, p_{14} is not able to treat the query that p_9 holds, thus such a query is not propagated further. Obviously, each friend would further spread the query, thus increasing the total number of relevant rewritings.

The following proposition holds.

Proposition 2. *If $|\mathcal{A}_Q|$ is the size (number of atoms) of an input query Q and $|M_r|$ the number of the relevant mappings in the PDMS then the number of rewritings generated by TranslateQuery is $O(|M_r|^{|\mathcal{A}_Q|})$.*

The upper bound of the above proposition is essentially due to the recursive callings triggered by the second loop (lines 10-30) and the third loop (lines 33-34), while the first loop (lines 6-8) is negligible. Also notice that the second loop only triggers one recursive call per mapping $\mathcal{M}_i$, according to the body (or head) relevance of the query Q with respect to the current mapping $\mathcal{M}_i$.

Algorithm 3: **FindDirectFOAFFriends** - Finds the top-k relevant "Simple Friends" and adds their entries in the FOAF file

---

Input : A query Q as set of atoms $\mathcal{A}_Q$ and a peer p with its list $LSVList$ of mappings $\Sigma_{1 \leq i \leq n} \mathcal{M}_i$ in the peer's local semantic view (LSV) and a FOAF file f
Output: The updated FOAF file f
1   Let L be a list of mappings ordered by relevance;
2   **foreach** mapping $\mathcal{M}_i$ in $LSVList$ **do**
3   RV = Call **ComputeRelevance**$(Q, \mathcal{M}_i.\text{MappingRules}())$;
4      $mapscore[i]$ = SumValuesFromVector(RV);

5   L = **Order** $LSVList$ according to mapping relevance values in $mapscore$;
6   **foreach** top-k mapping $\mathcal{M}_i$ in the ordered list L **do**
7   Let p' the target peer through $\mathcal{M}_i$;
8   **if** p' is not in the FOAF f **then**
9   Call **InvitePeer**(p, p');
           //Asynchronous method
10   **if** the previous invitation has been accepted **then**
11   Insert p' in the FOAF file f;

12   **return** the updated FOAF file f;

4.3 Seeking New Friends

Algorithm 3 updates the FOAF file of a given peer, by adding new friends, discovered after an exhaustive inspection of the content of the local semantic view of a peer. Before adding a peer p' to the FOAF file of the current peer, a formal invitation is sent and must be accepted. A simple extension of Algorithm 3 can be thought, in which an external peer, which is not friend of a friend, is added to the FOAF file. We omit its pseudocode for the sake of conciseness.

4.4 Gossiping Mapping Entries

To conclude this section, we discuss the gossip behavior of each peer. An active thread describes how a peer p initiates a periodic gossip exchange, while the passive thread takes care of a gossip exchange initiated by some other content peer p''.

The active behavior is triggered after each time interval $TGossip$. After incrementing the age of its view entries by 1, the peer p selects from its view: (a) a peer p', being the oldest contact via select_oldest() and (b) a viewSubset, being a random subset of Mapping-content within the local semantic view of $LGossip$ size (where $0 < LGossip <= VGossip$). Then, peer p send to p' a gossip message, a message that contains the viewSubset. Recall that each peer keeps in its LSV a set of the mappings containing a specific atom (see Fig. 4).

The peer p receives in exchange $gossipMsg'$ containing similar information from p', and creates a viewEntry related to p', with age 0. Next, peer p discards duplicates view entries through using Merge. This lets taking care of the problem of redundant rewriting sequences.

The passive behavior is triggered when peer p receives a gossip message containing Mapping-content and view entries from some peer p''. Peer p answers by sending back a gossip message with its own Mapping-content and view information, and updates its local view with via merge() and select_recent(), and finally updates the local Mapping-content with respect to the new view as described previously.

We omit the pseudocode of the Gossiping mapping entries Algorithm for lack of space.

5   Experimental Evaluation

We first describe in Sect. 5.1 the system setup. In Sect. 5.2, we assess the quality and efficiency of our rewriting technique, also with respect to traditional query reformulation approaches. We then focus on the scalability of our algorithms and their robustness with respect to network churn in Sect. 5.3.

5.1   Experimental Setup

Dataset and mapping generation. We have conducted our evaluation in PeerSim [25], an open source simulator for P2P protocols. In order to tweak our system at best, we implemented a pseudo-randomized generator of relational schemas. Indeed, none of the available relational datasets could provide us enough heterogeneity to distribute on a large number of nodes in the network. Thanks to this generator, no peer's schema is identical to any other and, as a consequence, mappings are all distinct. Moreover, every peer has at least one acquaintance, connected to it via a mapping. This ensures that there are not semantically disconnected peers in the PDMS.

The generator leverages a dictionary of about 40 names, ranging from table names to attribute names. We have designed a total of 10 scenarios (outlined in Table 1), by varying the number of peers in the PDMS, the number of mappings and the number of acquaintances, the latter ranging from a minimum of 1 to a 21, which is compatible with the diversity of the randomized schemas.

In the above scenarios, the number of mappings from a peer to each of its acquaintances ranges from 1 to 6, whereas each mapping has at most 3 atoms in the body/head. Moreover, each peer's schema has been randomly generated to contain at most 6 tables with at most 3 attributes each. The queries used in the experiments have been randomly generated to match the atoms in the body/head

Table 1. Heterogeneous scenarios used for experiments.

Scen.	# of Peers	# of Mappings	#Min Acq.	#Max Acq.
1	500	2767	3	12
2	1000	6202	2	14
3	1500	9941	2	15
4	2000	13814	2	16
5	2500	17893	2	17
6	3000	22037	1	18
7	3500	26394	1	18
8	4000	30696	1	20
9	4500	34941	1	21
10	5000	39261	1	21

of the mappings, thus may contain in turn from 1 to 3 atoms. Finally, the FOAF files are initially empty in all experiments, and are incrementally filled, as soon as query reformulation starts.

Qualitative measures and protocols for comparison. In each of the scenario depicted in Table 1, one or more queries are formulated on initiating peers and they fire a certain number of *relevant rewritings*, which represent all the rewritings for which the AF-IMF measure is greater than 0. To evaluate the quality of the top-k mappings, we run our query reformulation algorithms in a centralized implementation of our protocol, and take the returned results for each query as relevant rewritings. We have measured the *recall*, which is computed as follows:

$$\text{Recall}_{\text{AF-IMF}} = \frac{Number of Retrieved Relevant Rewritings}{Total Number of Relevant Rewritings}$$

Moreover, we have measured the time (in ms) taken to retrieve such relevant rewritings.

In order to gauge the effectiveness of our techniques and also to provide a yardstick for comparison, we have implemented the following protocols, that have been used throughout the experimental assessment:

Full The query gets translated against the relevant (using AF-IMF) rewriting sequences, by exploiting LSV, gossiping and FOAF links.

Full- The query gets translated against the relevant (using AF-IMF) rewriting sequences, by exploiting LSV, gossiping (i.e. the protocol Full without FOAF links).

Baseline# The query gets translated against the relevant (using AF only) rewriting sequences.

Baseline+ The original query gets translated against the mappings found in the traversal, and all its rewritings (relevant and non relevant) get propagated.

Baseline The original query gets translated against the mappings found in the traversal, and gets propagated as it is (i.e. rewritings are not propagated).

With Baseline and Baseline+, we have reimplemented the propagation strategy of existing approaches [6,16], adopting, however, the bidirectional translation semantics of our system.

Initial System Setup. We have executed an initial set of experiments, aiming at determining the gossip thresholds *VGossip* and *LGossip*. The former indicated the size of the Mapping-content table in the LSV, while the latter allows to control the size of a gossip message within each gossip cycle. Both parameters directly impact the effectiveness of the gossip protocol, since they indicate of what size an LSV and its buffer should be to harvest the highest number of relevant content in the network.

From the experiments, that we omit for conciseness, we observe that a *VGossip* size of 500 entries is a good trade-off between number relevant rewritings retrieved and time, while varying the gossip cycles from 1 to 10. We also observe that, if we keep *LGossip* of the same size as *VGossip* (in other words, we disseminate the entire LSV in gossip messages) or smaller, the results in terms of rewritings are not affected much. Therefore, we opted for a value of $LGossip = 100$ throughout the analysis.

Moreover, as the ranking function to use in the TranslateQuery algorithm, we have adopted the harmonic mean, which overcomes by 0.5 % the other ranking functions (averaged on 10 gossip cycles).

Finally, we have also empirically determined the maximum number of relevant requests $REQS$. We observed in a batch of initial experiments that the number of rewritings is affected by a value of $REQS$ greater than 0 only during the initial gossip cycles, whereas $REQS = 0$ becomes the most preferable choice, when the number of gossip cycles increases. From these experiments, we could infer that $REQS$ should be used as a dynamic threshold, and should have values slightly greater than 0 when gossiping starts and drop to 0 as long as gossip cycles reach 4.

Also, we have set the threshold α of the number of query hops to *unbounded*, to be able to observe the behavior of our algorithms in the most general case. Our prototype has been implemented in Java and all experiments have been performed on a 2.7 Ghz Intel Corei5 machine with 4GB RAM, running Windows 7 and JDK 6. For all experiments (unless otherwise specified), we have used a PDMS of 2000 peers with a configuration as in scenario 4 of Table 1.

5.2   Qualitative Evaluation

Recall and Comparison with previous approaches. As described above, we have measured the recall of our approach and compared it with the protocols *Baseline* and *Baseline+*. From Fig. 7 (a), we can observe that our protocol Full has the greatest recall along all the values of top-k mappings, if compared with all other protocols. In particular, the contribution of FOAF links to the recall is noticeable, since such links enable to connect network areas which would be otherwise unexplored in the query translation process, as shown by the trend of Full and Full-. Baseline, Baseline+ and Baseline# have a lower recall, as they do not exploit the relevance measure AF-IMF, thus the mappings that they exploit during query translation are in most cases not relevant. As long as more mappings are traversed, their recall improves, until Baseline# reaches the same recall of Full-, while it never reaches 100 % recall. The latter is only achieved by the Full protocol, by exploiting the FOAF links. Interestingly, this experiment showed the effectiveness of AF-IMF, LSV and gossiping (from Baseline up to Full-) and the utility of FOAF links (from Full- to Full). In particular, it can be observed that adopting a local metric for evaluating mappings (like the AF metric of the Baseline# protocol) works better than using no metric at all (Baseline and Baseline+). However, it performs worse than using the AF-IMF global metric (like in Full- and Full), which has the most desirable behavior amongst all scenarios.

Similarly, in terms of the number of relevant rewritings, as shown in Fig. 7 (b), the Full protocol is the one that can harvest the highest number at any value of the top-k mappings. Finally, we have quantified the cost incurred by the Full and Full- protocols with respect to the Baseline protocols. The results are reported

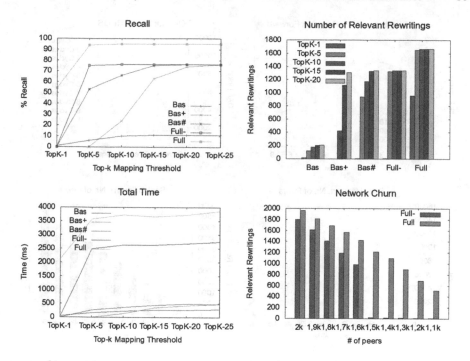

Fig. 7. (a) Recall, (b) # of Relevant Rewritings, (c) Total Time (ms) and (d) Network Churn from 2000 peers down to 1100 peers; 10 queries.

in Fig. 7 (c), which reports the time averaged on 10 queries. We can observe that the times undertake a certain increase, due mainly to the gossiping active and passive threads, and to the computation of relevance for Full- and, additionally, to the FOAF linking for Full. However, these times are still reasonable as the latter protocols allow a significant increase of the recall (as shown in Fig. 7 (a)).

Precision of distributed IMF. Next, we conducted another experiment to gauge the effectiveness of our distributed technique to compute AF-IMF. We have defined the precision of distributed IMF as follows:

$$\text{Precision}_{\text{IMF}} = \frac{ComputedIMFValue}{ExpectedIMFValue}$$

We recall that IMF only depends on the query and not on the mappings, whereas AF depends on the mappings. There are no false positives in the query reformulation process, therefore the precision of AF cannot be determined. For such a reason, the precision we have measured is defined on IMF.

By measuring such precision while varying the gossip cycles, we indirectly measure the effectiveness of the LSV. We can observe that in about 3 gossip cycles, the number of inquiries converges to $REQS = 0$, meaning that the LSV has fetched enough relevant tuples from the outside and is self-contained. The *backward* precision has a similar trend, and is omitted for lack of space.

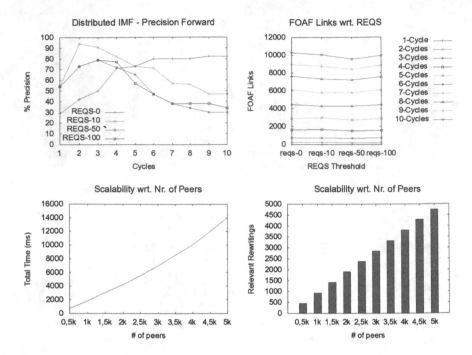

Fig. 8. (a) Distributed IMF Relevance Forward Precision, (b) Impact of $REQS$ on FOAF Links, (c) Scalability wrt. # of Peers (Time) and (d) Scalability wrt. # of Peers (Nr. of Rewritings); 10 queries.

Effectiveness of FOAF links. Figure 8 (b) shows another experiment we have conducted to gauge the increase of the number of FOAF links as the number of gossip cycles grows. Such increase is not affected much by the threshold $REQS$, thus confirming that the converging value of $REQS = 0$ already conveys enough FOAF links.

5.3   Scalability and Churn

We now assess the robustness of our techniques in large-scale PDMS, by varying the number of peers (spanning all scenarios in Table 1). In Fig. 8 (c) and (d), both the average time and # of relevant rewritings have a linear growth as the number of peers increases. This confirms that our techniques are scalable.

In the next experiment, we have simulated the network churn, by starting from an initial configuration of 2000 peers, and forcing 100 peers at a time to leave the network. The aim of this experiment was twofold, to measure the robustness of our PDMS to churn, and to show the utility of FOAF links in a situation in which acquantainces of the peers (along with their mappings) quit the network.

Figure 7 (d) shows that the Full- protocol (without FOAF) gets a few relevant rewritings after a cutoff point, i.e. when the # of peers drops to 1500; indeed, the useful acquaintances have left, and no FOAF link can be exploited to get new rewritings. Conversely, the Full protocol scales gracefully as the # of peers decreases and exhibit a linearly decreasing number of rewritings, thus showing the utility of FOAF links.

6   Related Work

There has been a great deal of work on data management in P2P databases on issues ranging from schema mediation [16] to mapping data values [20], query processing [22] and query translation [6,16].

Kementsietsidis et al. [20] describes a set of algorithms for exchanging data among peers, by only leveraging constraints on such exchange under the form of mapping tables, that comprise data values of the local peer and of external peers. Constraints on the content of peers under the form of logical rules are also studied in theoretical work on data integration [23].

The only previous work that considered query reformulation in this context is [6,16]. In Piazza [16], each peer is equipped with inclusion and equality mappings and a set of local storage descriptions. Query answering is done by evaluating the containment of any arbitrary external conjunctive query against the mappings and the storage descriptions. However, no approximation of the local peer mappings with suitable data structures is adopted. Moreover, [16] relies on a centralized index rather than on a distributed one. A schema mapping and query translation framework for XML databases is presented in both [6,29], which disregards the problem of ranking mappings based on relevance, as we do in this paper.

We focus on individual rewritings in this work and adopt the query rewriting semantics of [6,14]. Query rewriting with respect to a set of views is addressed in Minicon [27], where views are joined to return the maximally contained rewritings for LAV data integration.

Data integration in the presence of a global mediated ontology, relational data sources and GAV mappings is also addressed in [10].

Efficient XML query processing in P2P [22], leveraged multi-level Bloom-filters. However, we are not focusing on query optimization for XML, thus they are not directly comparable to our approach. Koloniari and Pitoura [22] study the problem of content-based routing by multi-level Bloom-filters, that works on XML data. However, they address XPath queries, and do not consider query reformulation.

Finally, Kantere et al. [19] consider the problem of clustering peers based on their common interests in unstructured networks. Contrarily to our approach, they utilize metrics to compare a query and its rewritings, that are applied after the rewritings have been computed and not beforehand, as in our approach. Moreover, our global AF-IMF metric is the first to take into account the entire collection of mappings in the network. Furthermore, we leverage individual mappings components, i.e. the atoms, to identify the most relevant mappings in the network, and assume that mappings are expressed as logical formulas (tgds), rather than being queries. Our definition of relevance relies on this assumption. The idea of quantifying the information transfer of individual schema mappings with local metrics is the subject of recent work [2]. However, no global metrics in a social and distributed context are considered. Although the previous metrics can be considered local, it would be interesting to see how they can combined with AF-IMF and deployed in a large-scale scenario, such as PDMS.

Gossiping [21] has been used for P2P network maintenance, and information dissemination.

A recent work [28] studies a collaborative tagging system, in which gossiping helps to personalize query processing, while computing the proximity between users tagging profiles. Being a collaborative tagging system, it deals with a different problem other than semantic heterogeneity. Gossiping as a mean to enter diverse semantic domains is used in [1], where basically mappings between peers may not be correct or simply not be aligned with a given domain. Therefore, the paper shows how local mappings can be used to establish a global semantic agreement among the peers.

7 Conclusions and Future Work

To conclude, in this paper we have studied the problem of reformulating conjunctive queries (CQs) in peer-to-peer data management systems by adopting the notion of social relevance. We have presented a new notion of relevance of a query with respect to a mapping, thus introducing a novel metric to rank mappings which have to be considered in the query translation process. By means of an extensive experimental analysis, we have proven the robustness of our approach by also exploiting the FOAF social network.

Our approach assumes the existence of mappings among peers but we do not require symmetric mappings (i.e. the existence of inverted mappings). Moreover, such mappings are currently settled among peers' relational schemas but we plan to extend our approach to both nested relational schemas and OWL ontologies in order to catch more semantics in the query reformulation process.

We also plan to cope with some limitations of our *backward relevance* assumption. Indeed, we currently consider a query Q to be *backward relevant* with respect to a mapping m if, and only if, each atom appearing in the query Q can be unified with an atom only containing universally quantified variables in the head of the mapping m. In the future work we will address this limitation by considering the adoption of unknown markers so that we will able to also capture those rewritings that, however, are relevant in a sense of information gain.

Finally, we will focus on studying the impact of query personalization, the combination with other quality metrics (both globals and locals) and the extension to unions of conjunctive queries (UCQs) also keeping into account peers' schemas with relational constraints.

References

1. Aberer, K., Cudré-Mauroux, P., Hauswirth, M.: The chatty web: emergent semantics through gossiping. In: WWW (2003)
2. Arenas, M., Pérez J, Reutter, J.L., Riveros, C.: Foundations of schema mapping management. In: *PODS*, pp. 227–238. ACM, New York (2010)
3. Arenas, M., PTrez, J., Riveros, C.: The Recovery of a Schema Mapping: Bringing Exchanged Data Back. In: ACM PODS, pp. 13–22 (2008)
4. Bizer, C., Heath, T., Berners-Lee, T.: Linked data - the story so far. Int. J. Semantic Web Inf. Syst. **5**(3), 1–22 (2009)
5. Bloom, B.H.: Space/time trade-offs in hash coding with allowable errors. Commun. ACM **13**(7), 422–426 (1970)
6. Bonifati, A., Chang, E.Q., Ho, T., Lakshmanan, L.V.S., Pottinger, R., Chung, Y.: Schema mapping and query translation in heterogeneous P2P XML databases. VLDB J. **19**(2), 231–256 (2010)
7. Bonifati, A., Chrysanthis, P.K., Ouksel, A.M., Sattler, K.: Distributed databases and peer-to-peer databases: past and present. SIGMOD Record **37**(1), 5–11 (2008)
8. Bonifati, A., Liu, R., Wang, H(.W).: Distributed and secure access control in P2P databases. In: Foresti, S., Jajodia, S. (eds.) Data and Applications Security and Privacy XXIV. LNCS, vol. 6166, pp. 113–129. Springer, Heidelberg (2010)
9. Bonifati, A., Mecca, G., Pappalardo, A., Raunich, S., Summa, G.: Schema mapping verification: the spicy way. In: EDBT, pp. 85–96 (2008)
10. Calvanese, D., De Giacomo, G., Lembo, D., Lenzerini, M., Poggi, A., Rosati, R., Ruzzi, M.: Data integration through $DL - Lite_A$ ontologies. In: Schewe, K.-D., Thalheim, B. (eds.) SDKB 2008. LNCS, vol. 4925, pp. 26–47. Springer, Heidelberg (2008)
11. Fagin, R.: Inverting Schema Mappings. ACM TODS **32**(4), 25:1–25:53 (2007)
12. Fagin, R., Haas, L.M., Hernández, M., Miller, R., Popa, L., Velegrakis, Y.: Clio: schema mapping creation and data exchange. In: Borgida, A.T., Chaudhri, V.K., Giorgini, P., Yu, E.S. (eds.) Conceptual Modeling: Foundations and Applications. LNCS, vol. 5600, pp. 198–236. Springer, Heidelberg (2009)
13. Fagin, R., Kolaitis, P.G., Miller, R.J., Popa, L.: Data exchange: semantics and query answering. TCS **336**(1), 89–124 (2005)
14. Fagin, R., Kolaitis, P.G., Miller, R.J., Popa, L.: Data exchange: semantics and query answering. Theor. Comput. Sci. **336**(1), 89–124 (2005)
15. Fan, L., Cao, P., Almeida, J.M., Broder, A.Z.: Summary cache: a scalable wide-area web cache sharing protocol. In: SIGCOMM, pp. 254–265 (1998)
16. Halevy, A.Y., Ives, Z.G., Suciu, D., Tatarinov, I.: Schema mediation for large-scale semantic data sharing. VLDB J. **14**(1), 68–83 (2005)
17. Hose, K., Roth, A., Zeitz, A., Sattler, K., Naumann, F.: A research agenda for query processing in large-scale peer data management systems. Inf. Syst. **33**(7–8), 597–610 (2008)
18. Ives, Z.G., Halevy, A.Y., Mork, P., Tatarinov, I.: Piazza: mediation and integration infrastructure for Semantic Web data. J. Web Sem. **1**(2), 155–175 (2004)
19. Kantere, V., Tsoumakos, D., Sellis, T.K., Roussopoulos, N.: GrouPeer: Dynamic clustering of P2P databases. Inf. Syst. **34**(1), 62–86 (2009)
20. Kementsietsidis, A., Arenas, M., Miller, R.J.: Mapping data in peer-to-peer systems: semantics and algorithmic issues. In: SIGMOD, pp. 325–336 (2003)
21. Kermarrec, A., van Steen, M.: Gossiping in distributed systems. Operating Systems Review **41**(5), 2–7 (2007)

22. Koloniari, G., Pitoura, E.: Content-based routing of path queries in peer-to-peer systems. In: Bertino, E., Christodoulakis, S., Plexousakis, D., Christophides, V., Koubarakis, M., Böhm, K. (eds.) EDBT 2004. LNCS, vol. 2992, pp. 29–47. Springer, Heidelberg (2004)
23. Lenzerini, M.: Data integration: a theoretical perspective. In: ACM PODS, pp. 233–246 (2002)
24. Levy, A.Y., Mendelzon, A.O., Sagiv, Y., Srivastava, D.: Answering queries using views. In: PODS (1995)
25. The Peersim simulator. http://peersim.sf.net
26. Popa, L., Velegrakis, Y., Miller, R.J., Hernandez, M.A., Fagin, R.: Translating web data. In: VLDB (2002)
27. Pottinger, R., Halevy, A.Y.: Minicon: a scalable algorithm for answering queries using views. VLDB J. 10(2–3), 182–198 (2001)
28. Bai X., Bertier, M., Guerraoui, R., Kermarrec, A.M., Leroy, V.: Gossiping personalized queries. In: EDBT, pp. 87–98 (2010)
29. Yu, C., Popa, L.: Constraint-based XML query rewriting for data integration. In: SIGMOD (2004)

Distributed Large-Scale Information Filtering

Christos Tryfonopoulos[1]([⊠]), Stratos Idreos[2], Manolis Koubarakis[3],
and Paraskevi Raftopoulou[1]

[1] University of Peloponnese, Tripoli, Greece
{trifon,praftop}@uop.gr
[2] Harvard University, Cambridge, MA, USA
stratos@seas.harvard.edu
[3] National and Kapodistrian University of Athens, Athens, Greece
koubarak@di.uoa.gr

Abstract. We study the problem of distributed resource sharing in peer-to-peer networks and focus on the problem of information filtering. In our setting, subscriptions and publications are specified using an expressive attribute-value representation that supports both the Boolean and Vector Space models. We use an extension of the distributed hash table Chord to organise the nodes and store user subscriptions, and utilise efficient publication protocols that keep the network traffic and latency low at filtering time. To verify our approach, we evaluate the proposed protocols experimentally using thousands of nodes, millions of user subscriptions, and two different real-life corpora. We also study three important facets of the load-balancing problem in such a scenario and present a novel algorithm that manages to distribute the load evenly among the nodes. Our results show that the designed protocols are scalable and efficient: they achieve expressive information filtering functionality with low message traffic and latency.

Keywords: Publish/subscribe · Distributed hash tables · Information management

1 Introduction

Peer-to-peer (P2P) computing has been around for more than a decade, contributing a vast amount of research results and deployed prototypes. In this context applications that scale to millions of users and resources have been developed in domains like file-sharing (e.g., bitTorrent and Kazaa), voice and video distribution (e.g., Skype and P2PTV), distributed search engines (e.g., FAROO, YaCy, and the scientific search engine Sciencenet), and even project management (e.g., Collanos workplace).

Apart from the traditional applications discussed above, the techniques, tools and architectures introduced by the P2P paradigm have found new, interesting

Part of this work was performed while the authors were with the Technical University of Crete, Chania, Greece. C. Tryfonopoulos was partially supported by programme Heraclitus of the Greek Ministry of Education.

A. Hameurlain et al. (Eds.): TLDKS XIII, LNCS 8420, pp. 91–122, 2014.
DOI: 10.1007/978-3-642-54426-2_4, © Springer-Verlag Berlin Heidelberg 2014

and useful applications also in other domains. Lately, P2P concepts are also being introduced in the –clearly different– cloud paradigm (e.g., in P2P-assisted cloud provisioning [1–3], or in hybrid architectures [4] with a backbone of super-peers that provides access to cloud users and serves queries by executing distributed protocols), in an effort to exploit the benefits of both technologies. Additionally, P2P architectures and protocols have also been proposed in the context of distributed online social networking [5–7], aiming at solving content ownership and scalability issues, while minimising deployment and maintenance costs.

In this article, we present P2P protocols to support content and/or service lookup by utilising structured overlays, aiming at content-based filtering functionality in distributed environments.

Targeted Functionality. In the P2P architecture that we envision, resources are annotated using attribute-value pairs, where value is of type text and queried using constructs from Information Retrieval (IR) models. There are two kinds of basic functionality that we expect this architecture to offer: *information retrieval* and *information filtering (IF)*. In an IR scenario a user poses an one-time *query* and the system returns a list of pointers to matching resources. In an IF scenario, also known as *publish/subscribe (pub/sub)*, a user posts a *subscription* (or *profile* or *continuous query*) to the system to receive notifications whenever certain events of interest take place. In this article, we concentrate on the latter kind of functionality (i.e., IF) and show how to provide it by extending the distributed hash table Chord [8]. We assume that publications and subscriptions will be expressed using a well-understood attribute-value model, called $\mathcal{AWPS}$ [9]. $\mathcal{AWPS}$ is based on *named attributes* with value *free text* interpreted under the Boolean and vector space (VSM), or latent semantic indexing (LSI) models.

Our architecture and protocols target dynamic information dissemination applications, such as news alerts, digital libraries, weather monitoring, and stock quotes, consisting of large, open, and dynamic user communities. Especially for cases like news alerts and digital libraries –where the data of interest is mostly textual and users express their needs using IR languages (for example, keywords or pieces of text)– our architecture is well-suited as an implementation technology. It can handle huge amounts of information in a highly distributed, self-organising way, while offering benefits in terms of openness, scalability, and efficiency. Following our approach users, or services that act on users' behalf, would specify continuous queries for information, thus subscribing to newly appearing documents that satisfy the query conditions. The system would then notify the subscribed users automatically whenever a new matching document is published. Publishers in such a setting could be news feeds, digital libraries, or users, who post new items to blogs or other Internet communities.

Contributions. The contributions of this article are the following. We present a set of *novel protocols*, collectively called *DHTrie*, that extend the Chord protocols with pub/sub functionality assuming that publications and subscriptions are expressed in the model $\mathcal{AWPS}$. In a distributed pub/sub environment, publications typically involve contacting a large set of nodes, where matching with stored subscriptions takes place. To do this effectively, we have designed and implemented four methods that target low network traffic and low latency. In combination with these methods, we introduce a simple routing table that uses only local information and manages to significantly reduce network traffic. To justify our solution, we evaluate the DHTrie protocols experimentally in a distributed digital library scenario with hundreds of thousands of nodes and millions of user profiles. Our experiments show that the DHTrie protocols are *scalable*: the number of messages needed to publish a document and notify interested subscribers remains almost constant as the network grows, while latency is kept low. As probability distributions associated with words in publications and queries are skewed, balancing the node load becomes an important issue. We study three cases of load balancing for DHTrie, namely *query*, *routing* and *filtering* load balancing, and present a new algorithm that tackles the load balancing issues.

Preliminary results of this research have appeared in [10]. The current article revises [10] and presents the following extensions and additional contributions. We consider the issue of latency in addition to that of network traffic and identify the relevant tradeoff in our experimental evaluation. To tackle this tradeoff, we introduce two novel methods (called *hybrid* and *continuous splitting*) for resource publication. The new approaches, although very different in philosophy and design, manage to keep publication latency low while performing well in terms of network traffic. The hybrid method is a family of novel tunable alternatives that allow a per-node parameter setting aiming at adaptability. The continuous splitting method is automatic and parameterless, which makes it easy to deploy; it has proven to be efficient to many different settings and goals.

In addition to the above novel contributions, we also include more detailed descriptions of the DHTrie protocols and their respective data structures (see Sect. 3), and extend the experimental work (Sect. 4) with measurements of the new methods and comparison with the ones presented in [10], comparison under two different corpora, and experiments for publication latency and network dynamics. Finally, we redesign and apply the algorithm presented in [10] to query load balancing, and study its effects on message traffic (Sect. 4.9).

Organisation. The organisation of the article is as follows. Section 2 positions our work with respect to related research, while Sect. 3 presents the DHTrie protocols. The experimental evaluation of DHTrie and a study of load balancing issues are presented in Sect. 4, followed by Sect. 5 that concludes the article.

2   Related Work and Background

In this section, we survey related work in the area of pub/sub and IF in P2P networks. Naturally, this paper is also relevant with the broad area of distributed query processing, with studies on different query models in distributed settings (i.e., one-time, relational, and RDF query processing), and with the area of IR in P2P networks as it shares many common goals and techniques with IF.

2.1   P2P Pub/Sub and Information Filtering

Work on pub/sub in distributed systems has contributed some fundamental ideas that have also been utilised in the P2P domain. Researchers in this area have developed various data models based on channels, topics, and attribute-value pairs to represent publications and subscriptions. Pub/sub systems based on attribute-value models are called *content-based*, as attribute-value data models are flexible enough to express the content of publications. The query languages of content-based pub/sub systems are based on Boolean combinations of arithmetic and string operations. Work in this area has concentrated not only on distributed pub/sub architectures, but also on filtering protocols.

SIENA [11] is probably the most elegant example of a system to be developed in this area. A very important contribution of SIENA is the adoption of a P2P model of interaction among servers and the exploitation of traditional network algorithms based on shortest paths and minimum-weight spanning trees for routing messages. The core ideas of SIENA have been used in the early P2P pub/sub system P2P-DIET [12].

With the advent of DHTs such as CAN, Chord, and Pastry a new wave of pub/sub systems has appeared. Scribe [13] is a topic-based pub/sub system based on Pastry. Hermes [14] is similar to Scribe since it uses the same underlying DHT but it allows more expressive subscriptions by supporting the notion of an event type with attributes.Related ideas appear later in [15,16] and in PeerCQ [17], a notable pub/sub system implemented on top of a DHT infrastructure designed to cope with peer heterogeneity by extending consistent hashing [18].

The study in [19] is mainly concerned with scalability of current designs and proposes two methods that allow to restrict the overall costs.Both these methods can improve general purpose P2P protocols and can be applied on top of our work as well. Triantafillou and Aekaterinidis [20] study the problem of content-based pub/sub functionality on top of Chord, allowing for range-based subscriptions, i.e., one can define a range for a given attribute as opposed to a single value.Such ideas can readily be adopted by our protocols as well. Meghdoot [21] is another pub/sub proposal in the area of DHTs that uses ideas such as hashing of subscriptions and events to facilitate matching.The difference of Meghdoot from our work is that it is built on top of CAN, whose characteristics are heavily exploited in the system design (e.g., it uses zone splitting/replication).

Research on processing subscriptions using string attributes in DHT-based pub/sub systems is also related to our work. PastryStrings [22] utilises prefix-based routing to facilitate processing of publications that are strings, and subscriptions that are string predicates. Additionally, the DHTStrings system [23] utilises a DHT-agnostic architecture to support prefix and suffix queries in string attributes. More recent works on P2P pub/sub systems have focused on various issues such as new routing protocols [24–26], combination of IR and IF [27], web services [28], load balancing [29], security [30] and preference awareness [31].

Similarly to the pub/sub strand of research, approaches that use a DHT as the routing infrastructure to build filtering functionality for IR-based models and languages have also been introduced. Closer to our work are the systems pFilter [32] and Ferry [33].The main qualitative difference of our work is that we support a different and more expressive query model, requiring more complex protocols. In addition, from a quantitative point of view our work provides a more in depth analysis by stressing the system to millions of queries and tens of thousands of nodes as opposed to only thousands of queries and thousands of nodes in [32,33]. Below we discuss these works in more detail, and compare them against our approach.

pFilter [32] is the closest system to the ideas presented in this work. It uses a hierarchical extension of CAN [34] to store user queries and relies on multi-cast trees to notify subscribers. Compared to pFilter, our work uses a more expressive data model and query language, while there is no need to maintain multi-cast trees to notify subscribers. However, the multi-cast trees of pFilter take into account physical network distance something that we do not consider at all in this work, but rather we consider publication latency and load balancing issues.

Ferry [33] is another proposal to support IF functionality on top of DHTs. The main difference of our work is the support of a more expressive and complex data and query model. The main novelty of Ferry is that it exploits the DHT links, e.g., the contents of the Chord finger table, to disseminate information in the network. In our work, we exploit similar ideas by extensively taking advantage of DHT links, trying to group messages based on the Chord finger table, and piggy-backing information on maintenance messages. In addition, we provide further routing flexibility with the addition of a routing table, called FCache (see Sect. 3.5), that consists of a low cost and best effort cache of IP addresses that allows us to bypass the DHT protocol whenever this is possible. Essentially, this comes at zero network cost as it is a process piggy-backed on the normal DHT messages.

2.2   Other Related Areas

Distributed query processing relies on distributed protocols that dictate where data meets queries. Depending on the network design and properties, and on the query model utilised, different query processing algorithms are needed. The first systems to cope with distributed query processing were mainly based on strictly structured designs and focused on relational query processing [35].

Mariposa [36], one of the most well-known distributed database systems and probably the most ambitious attempt to scale to thousands of nodes, proposes node interaction protocols based on economic models.Another well known distributed database system is LH* [37], where the authors introduce the notion of the scalable distributed data structure (SDDS).

The *early P2P designs* in the area tried to remove all the restrictions of the classic distributed systems.However, the more demanding nature of applications enforced structure in P2P networks, appearing in the form of DHTs and hypercubes [38]. Essentially, these architectures provide functionality so as data items or queries can be *mapped* to certain node(s) given a set of properties and functions.In this way, structured networks provide a non-centralised but still controllable design pattern.

These network designs can be seen as a hybrid between the early distributed systems and the early P2P networks, trying to balance the various tradeoffs and thus, offering extensive flexibility and adaptability to build any kind of application over it. Thus, there has been a tremendous amount of research over structured P2P networks, e.g., there is work on relational one-time and continuous query processing [17,39–43] and RDF query processing [44–47].

Information retrieval is the dual problem of information filtering, often referred to the other side of the same coin [48]. Although many of the underlying issues are similar as both IR and IF share the common goal of information delivery to information seekers, the design issues (e.g., timeliness of data, identification and representation of user needs), and also the techniques and protocols to satisfy these information needs differ significantly.

In [40], one of the early works that considered how to process IR queries on top of DHTs, the authors discuss issues involved in building IR functionality over text databases on top of structured overlays.In a similar spirit, [49] discusses the feasibility of Web search in a P2P environment and estimates the difficulty of the problem.A straightforward approach to support Boolean searching in P2P networks is presented in [50], where each node in the network is responsible for a specific keyword through the DHT hash function and the focus is put on multiple keyword queries.

Many works have studied how to support document querying based on VSM on top of structured overlays. Meteorograph [51], one of the early works that deal with similarity search over structured P2P networks, describes how to support similarity and ranked search in a linear hash addressing space overlay. In another approach, LibraRing [52] proposes a two-tier architecture for a digital library environment aiming to unify IR and IF in a single framework.While most of related papers utilise a DHT to route the queries to appropriate peers, in Minerva [53] a global distributed directory for IR-style statistics and quality of service information is built at indexing time, to be then exploited at query time.

Finally, *cloud/grid computing* and *social networking* have emerged over the last couple of years as new paradigms and application areas for distributed data management. Our research is also related to works in this domain, as researchers exploit and extend ideas from the distributed/P2P domain to provide new data management functionality as in [1, 5, 6, 54, 55].

3   The DHTrie Protocols

We implement pub/sub functionality by a set of protocols called *DHTrie* (from the words DHT and trie). The DHTrie protocols use *two levels of indexing* to store submitted queries.

The first level corresponds to the partitioning of the global query index to different nodes using DHTs as the underlying infrastructure. Each node is responsible for a fraction of the submitted queries through a mapping of attribute values to node identifiers. The DHT infrastructure is used to define the mapping scheme and also manages the routing of messages between different nodes. We use an extension of the Chord DHT [56] to implement our network. The set of protocols that regulate node interactions are described in the next sections.

The second level of our indexing mechanism is managed locally by each node and is used for indexing the user queries the node is responsible for. Each node uses a trie-like data structure to perform query clustering and improve filtering performance. The details of local indexing are presented in [57].

3.1   The Subscription Protocol

Let us assume that a node P wants to subscribe with a query q composed as a conjunction of atomic queries:

$$A_1 = s_1 \wedge \ \dots \ \wedge \ A_m = s_m \ \wedge$$
$$A_{m+1} \sqsupseteq wp_{m+1} \ \wedge \ \dots \ \wedge A_n \sqsupseteq wp_n \ \wedge \qquad (1)$$
$$A_{n+1} \sim_{a_{n+1}} s_{n+1} \ \wedge \ \dots \ \wedge \ A_k \sim_{a_k} s_k$$

where A_i is an attribute, s_i is a text value, wp_i is a conjunction of words and *proximity formulas*[1], and a_i is a *similarity threshold*, i.e., a real number in the interval $[0,1]$. For a query q of the above form, the atomic queries with equality ($=$) and containment ($\sqsupseteq$) operators will be called its *Boolean part*, while the atomic queries with similarity ($\sim$) operators will be called its *vector space part*.

To perform the subscription, P randomly selects a single word w contained in any of the text values $s_1, \dots, s_m$ or word patterns $wp_{m+1}, \dots, wp_n$ and computes $H(w)$, where $H()$ is a consistent hash function used to map identifiers in the identifier circle of Chord [56], to obtain the identifier of the node that will be responsible for query q. Then P creates message FwDQUERY($id(P), ip(P), qid(q), q$), where $id(P)$ is the identifier of node P computed by hashing a piece of information that identifies P (e.g., its IP address and port, or a unique identifier given to it the first time it joins the network), $ip(P)$ is the IP address of P, and $qid(q)$ is a unique query identifier assigned to q by P. This message is then forwarded in $O(logN)$ steps to the node with identifier $H(w)$. Since only one node has to be contacted, this forwarding is done using the Chord *lookup*() function to locate *successor*($H(w)$), i.e., the first node which is equal or follows $H(w)$ clockwise in the identifier space and is called the *successor* node of identifier $H(w)$. Once *successor*($H(w)$) is located, it is directly contacted by P. In this way, queries of this type are always indexed under their Boolean part to save message traffic, since they need to be stored at a single node. Notice also that both $id(P)$ and $ip(P)$ need to be sent to the node that will store the query to facilitate notification delivery.

When P wants to submit a query q of the form $A_{n+1} \sim_{a_1} s_1 \wedge \dots \wedge A_n \sim_{a_n} s_n$ (i.e., with a VSM part only), it sends q to *all* nodes in the list $L = \{H(w_j) : w_j \in D_1 \cup \dots \cup D_n\}$, where $D_1, \dots, D_n$ are the sets of *distinct* words in text values $s_1, \dots, s_n$. In contrast to queries with a Boolean part described above, queries with *only* a VSM part need to be stored in all the nodes involved (computed as above) in order to ensure correctness of the filtering process. Sending the same message to more than one recipients is discussed in detail in the next section, where the same problem is posed again by the publication forwarding process.

When a node P' receives a message FwDQUERY containing q, it stores q using the second level of our indexing mechanism. P' uses a hash table to index all the atomic queries of q using as key the attributes $A_1, \dots, A_k$. To index each atomic query, three different data structures are also used: (i) a hash table for text values $s_1, \dots, s_m$, (ii) a trie-like structure that exploits common words in word patterns $wp_{m+1}, \dots, wp_n$, and (iii) an inverted index for the most "significant" words in text values $s_{n+1}, \dots, s_k$. P' utilises these data structures at filtering time to find quickly all queries q that match an incoming publication p. This is done using a

---

[1] A proximity formula is an expression of the form $w_1 \prec_{\xi_1} \cdots \prec_{\xi_k} w_k$, where w_i is a word and ξ_i is a distance interval of the form $\{[l, u]: l, u \in \mathbb{N}, l \geq 0 \text{ and } l \leq u\} \cup \{[l, \infty): l \in \mathbb{N} \text{ and } l \geq 0\}$. The proximity operator $\prec_{\xi}$ is used to capture the concepts of *order* and *distance* between words in a text document using intervals that impose lower and upper bounds on distances between words.

method that combines algorithms BestFitTrie [57] and SQI [58]. The details of local storage and indexing using BestFitTrie are discussed thoroughly in [57].

3.2   The Publication Protocol

Publication of a resource involves sending the same message to a group of nodes that is not known a priori. To tackle this problem, we have designed and implemented four methods: (i) the iterative method, which is the standard way to contact a number of different nodes over Chord, (ii) the recursive method, which creates a single message with all the recipients contained in a sorted list and works its way around the identifier space until all recipients have been contacted, (iii) the hybrid method which uses machinery from the two previous methods to provide a tunable alternative between the two extremes, and (iv) the continuous splitting method, which exploits the finger tables of all message recipients to split the message at every forwarding node, aiming at the optimisation of network traffic and latency.

The publication protocol essentially involves sending the same message to the group of nodes that are responsible for the distinct words contained in the text values of the different attributes of p. In this way, when a node P wants to publish a resource, it first constructs a publication of the form $p = \{(A_1, s_1), (A_2, s_2), \ldots, (A_n, s_n)\}$ (i.e., a set of attribute-value pairs (A_i, s_i), where A_i is a *named attribute*, s_i is a *text* value, and all attributes are *distinct*) that is the resource description. Let $D_1, \ldots, D_n$ be the sets of *distinct* words in $s_1, \ldots, s_n$. Then, publication p has to be propagated to *all* nodes with identifiers in the list $L = \{H(w_j) : w_j \in D_1 \cup \cdots \cup D_n\}$. The subscription protocol guarantees that L is a superset of the set of identifiers responsible for queries that match p. To propagate publication p in the DHT, P removes duplicates from L and sorts it in ascending order clockwise starting from $id(P)$. In this way, we obtain at most as many identifiers as the distinct words in $D_1 \cup \cdots \cup D_n$, since a node may be responsible for more than one of the words contained in the document.

3.3   Methods for Subscription and Publication

In this section, we describe four different methods to implement the subscription and publication protocols and present their advantages and disadvantages.

The Iterative Method. Each node P, that uses the iterative method (It) to contact the recipients in list L, constructs a FwdResource($id(P), pid(p), p, id(P')$) message for each identifier $id(P')$ contained in L, where $pid(p)$ is a unique metadata identifier assigned to publication p by node P. Then, it utilises the $lookup()$ procedure provided by Chord to locate node P' and sends it the FwdResource message. This is repeated for all the identifiers in L in an *iterative* way. Using this method, P needs $O(h \log N)$ messages, where h is the number of different nodes to be contacted. Figure 1 illustrates graphically the publication of a resource to three recipients under Chord using the iterative method and shows a message graph for a general case of resource publication under It.

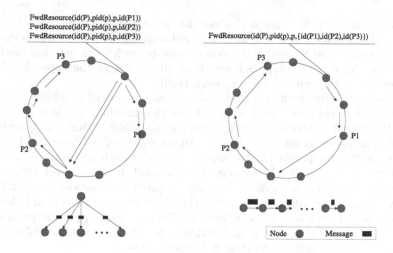

Fig. 1. Message routing and message graph for the *It* (left) and *Re* (right) methods

The Recursive Method. Using the iterative method has an obvious disadvantage; the same node may participate in many *lookup()* requests for nodes responsible for identifiers in list L causing increased network traffic. This is the reason for designing the recursive method (Re). The idea behind method Re is to pack messages together to reduce network traffic as follows.

Having obtained L, P creates a message FwdResource $(id(P), pid(p), p, L)$, where $pid(p)$ is a unique metadata identifier assigned to p by P, and sends it to node with identifier equal to $head(L)$ (the first element of L). This forwarding is done by the following *recursive* way: message FwdResource is sent to a node P', where $id(P')$ is the greatest identifier contained in the finger table of P, for which $id(P') \leq head(L)$ holds.

Upon reception of a message FwdResource by a node P', $head(L)$ is checked. If $id(P') < head(L)$ then P' just forwards the message as described in the previous paragraph. If $id(P') \geq head(L)$ then P' makes a copy of the message, since this means that P' is one of the intended recipients contained in list L (in other words P' is responsible for key $head(L)$). Subsequently, the publication part of this message is matched with the node's local query database using the algorithms described in detail in [57] and the appropriate subscribers are notified. Additionally, P' modifies list L to L' by deleting all elements of L that are smaller than $id(P')$ starting from $head(L)$, since all these elements have P' as their intended recipient. For the new list L', $id(P') < head(L')$ holds. Finally, P' forwards the message to node with identifier $head(L')$. Figure 1 illustrates graphically the publication of a resource to three recipients under Chord using the recursive method and shows a message graph for a general case of resource publication under Re.

The Hybrid Methods. The idea behind the recursive method is to pack messages together to reduce network traffic. This however, comes at the cost of high latency; if the recipients list is long then the last recipient has to wait for a long time until it is notified about the publication, which in turn causes delays in the notification of the interested subscribers. The iterative method on the other hand, tries to optimise latency since no recipients lists are used and the delay to deliver a message is logarithmic in the size of the network. This of course comes at the price of high network traffic.

To tackle this tradeoff, we designed and implemented a family of *hybrid* approaches that combine the benefits of the two previous methods. The idea behind the hybrid methods is to design tunable alternatives that will provide fast delivery of messages at low network cost. To achieve this, the message originator splits the initial recipients list to smaller ones and each recipients list is sent in an iterative way, while the message is forwarded in the network recursively. The family of hybrid methods is designed to provide variations with different objectives, while the difference between the three variants presented below lies in the initial splitting of the recipients list. Finally, notice that the parameters in the hybrid methods are not global, but may be set in a per-node fashion, thus adapting to node specifics regarding publication size.

The *fixHy* method. The fixed hybrid (*fixHy*) method requires fixing a value for the *desired recipients list size* σ. Parameter setting in the *fixHy* method, although adhoc, allows the manual tuning of the system according to document and vocabulary size, but requires expertise in setting this value. Notice that, if the average document length published in the system is changed, the method may create too short or even useless recipients lists. The *fixHy* method works as follows.

Having obtained L, node P uses it to create $h = \lceil |L|/\sigma \rceil$ recipients lists, of size σ. In our experiments, we used $\sigma = 10$ and $\sigma = 50$ as baseline values depending on the tested corpus, and showed the effect of the desired recipients list size in the message traffic and latency observed in the network. Notice that the *fixHy* method will degenerate to the recursive method for $\sigma = |L|$ and to the iterative method for $\sigma = 1$. Thus, using a high value for σ will make the protocol behave similarly to the recursive method, while using a low value for σ will make the protocol behave similarly to the iterative method.

The *perHy* method. The percentage hybrid (*perHy*) method requires the tuning of parameter π, which controls the percentage of the initial recipients list that will be used to create each new list. This method is less flexible than *fixHy* in setting the size of the recipients list, but requires less expertise and is adaptable to changes in the document size published in the network. Setting the recipient list size as a percentage of the initial recipients list allows coping with both large and small documents, whereas in the *fixHy* method this is not possible. The *perHy* method works as follows.

Having obtained L, node P uses it to create $h = \lceil 1/\pi \rceil$ recipients lists of size $|L| * \pi$, where $0 < \pi \leq 1$. In our experiments, we used $\pi = 4\%$ and also showed the effect of π in the message traffic and latency observed in the network. Notice that the *perHy* method will degenerate to the recursive method for $\pi = 1$ and to the iterative method for very small values of π.

The *medHy* method. The median hybrid (*medHy*) method is an automatic method that requires no parameter tuning, since the recipients lists are split according to the median of the differences between consecutive intended recipients. This method identifies the large "gaps" in the intended recipients list and splits it accordingly. Since no parameter setting is required, this is a method best suited for general purpose applications with published documents of varying size; no expertise is needed, since the recipients list is split according to its special characteristics. The *medHy* method works as follows.

Having obtained $L = \{l_1, l_2, \ldots, l_{|L|}\}$, node P traverses it starting at $head(L)$ = l_1 and calculates all differences $\delta_i = l_i - l_{i+1}, 1 \leq i \leq |L| - 1$, between consecutive intended recipients in L. Subsequently, P calculates the median δ_{med} of these differences and uses it to split L in the following way. P traverses L once more starting at $head(L)$, and when $\delta_k > \delta_{med}, 1 \leq k \leq |L| - 1$, it creates a new list $L_1 = \{l_1, \ldots, l_k\}$. Subsequently, L becomes $L \setminus L_1$, while element l_{k+1} is now $head(L)$, and the process continues until L is empty. Notice that there is no way to tune this method to behave similarly to either the iterative or the recursive method, since the splitting of the initial intended recipients list is done automatically according to the keys in L. The *medHy* method provides an automatic way to utilise the hybrid protocol, without the need for performance tuning, or any knowledge of the underlying document properties.

For each one of the lists $L_1, \ldots, L_h$ created by any of the variations ($fixHy$, $perHy$, $medHy$) of the hybrid method presented above, a message of the form FWDRESOURCE($id(P),pid(p),p,L_i$), with $1 \leq i \leq h$, is constructed and is iteratively sent to $head(L_i)$. Since each message contains a list of recipients, the recursive method is utilised to forward the message to the rest of the nodes in list L_i. When a node P' receives a FWDRESOURCE message, it removes all elements in L that have P' as their intended recipient and forwards the message in a recursive way. Notice that only the message originator may split the message into smaller lists, while the rest of the nodes receiving it are responsible just for forwarding it. The usage of many recipients lists with smaller size together with the iterative way of sending these lists justifies the hybrid nature of the protocol. As we will show in Sect. 4, this method manages to achieve lower latencies than the recursive method while keeping message traffic relatively low. Figure 2 illustrates graphically the publication of a resource to three recipients under Chord using any of the hybrid methods and shows a message graph for a general case of resource publication under $fixHy$, $perHy$, or $medHy$.

The Continuous Splitting Method. All the variations of the hybrid method split the initial recipients list only once at the message originator and then, all the subsequent recipients of the messages are forced to perform the routing task based on these lists. This makes the protocol simpler, but also adds inefficiencies, since the recipients of the messages are not allowed to optimise the routing by splitting the message further. The continuous splitting (Spl) method overcomes this limitation by allowing each message recipient to split the message into sublists, according to information in its finger table. In this way, the message is split several times at each recipient before it is forwarded to other intended recipients, causing an adaptive execution of the forwarding process.

The Spl method tries to exploit each node's view of the network (in contrast to the hybrid methods that exploit only the originators' view), by splitting the intended recipients lists according to finger table entries of all the nodes participating at the forwarding of the message. The drawbacks of this method

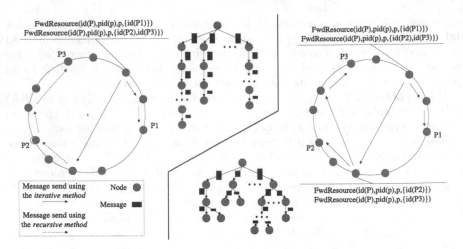

Fig. 2. Message routing and message graph for the $fixHy/perHy/medHy$ (left) and Spl (right) methods

include the need to access the finger table of each node, which may result in poor splitting of the intended recipients list if the finger table entries are outdated, or the message is too small. The Spl method works as follows.

Each finger table entry $f_P[i]$ in the finger table of P is used to create one or more lists in the following way. Consider two consecutive entries in the finger table of P, say $f_P[i]$ and $f_P[i+1]$. Starting from the identifier $id(f_P[i])$ stored at this entry, P scans $L = \{l_1, l_2, \ldots, l_{|L|}\}$, and collects all the recipients with identifier greater than $id(f_P[i])$ and smaller than $id(f_P[i+1])$ to create list $L_1 = \{l_1, \ldots, l_k\}, 1 \le k \le |L| - 1$. Subsequently, L becomes $L \setminus L_1$, while element l_{k+1} is now $head(L)$ and the process continues for all the intended recipients in the list L, until L is empty. Typically, finger entries with higher index number have longer lists associated with them (remember that entries in the finger table of a node point to exponentially increasing distances away from the node), which means that typically the distance between the identifiers of entries $f_P[i-1]$ and $f_P[i]$ is shorter than the distance between those of entries $f_P[i]$ and $f_P[i+1]$.

For each one of the lists $L_1, \ldots, L_h$ created by the continuous splitting method, a message FWDRESOURCE($id(P), pid(p), p, L_i$), with $1 \le i \le h$, is constructed and is iteratively sent to $head(L_i)$. When a node P' receives a FWDRESOURCE message, it removes all elements in L that have P' as their intended recipient and repeats the procedure described above to split list L_i further according to its own finger table. As we will show in Sect. 4, this method manages to achieve latency as low as that of the iterative method while keeping message traffic low. Figure 2 illustrates graphically the publication of a resource to three recipients under Chord using the continuous splitting method and shows a message graph for a general case of resource publication under Spl.

The reader may have noticed that the publication (and also the subscription) protocol in all the proposed methods indexes queries that consist of a single equality of the form $A = s$ using a single word contained in the text value s, contrary to the standard way that would index the entire text value s in the DHT. This is done to avoid sending extra network messages for each publication to discover matching equalities. False positives that may occur are resolved locally at each node, thus relieving the network of significant message overhead.

Independently of [10], where we originally presented the iterative and recursive methods, the technical report [59] presented an approach that shares ideas with these two methods by discussing how to implement multicast functionality at different levels of a DHT architecture. However, [59] aimed at multicast from a physical network viewpoint and focused on the comparison of these techniques across the CAN and Chord DHTs.

3.4 The Notification Protocol

When a message FWDRESOURCE containing a publication p arrives at a node P, the queries matching p are found by utilising its local index structures and using the algorithms described in detail in [57] for queries with a Boolean part only. The extension to $\mathcal{AWPS}$ queries involves the calculation of the cosine of the angle of two vectors corresponding to text values from a publication and a query, and follows straight-forward IR techniques.

Once all the matching queries have been retrieved from the database, P creates notification messages of the form NOTIFICATION$(ip(P), pid(p), qid(q))$, where P is the provider that published the matching resource, and sends them to all the nodes that their queries were matched against p using their IP addresses associated with the query they submitted. If a node P' is not online when P tries to notify it about the published resource, the notification message is sent to the $successor(P')$. In this way P' will be notified the next time it logs on the network. To utilise the network in a more efficient way, notifications can also be batched and sent to the subscribers when traffic is expected to be low.

3.5 Frequency Cache

In this section, we introduce an additional routing table that is maintained in each node. This table, called *frequency cache (FCache)*, is used to reduce the cost of publishing a resource. Using the protocols described earlier, each node is responsible for handling queries that contain a specific word. When a resource r with h distinct words is published by node P, P needs to contact at most h other nodes which will match the incoming resource against their local query databases. This procedure costs $O(h \log N)$ messages for each resource published at P. Since some of the words will be used more often at published resources, it is useful to store the IP addresses of the nodes that are responsible for queries containing these words. This allows P to reach in a single hop the nodes that are contacted more often (proxying).

Specifically, FCache is a hash table used to associate each word that appears in a published document with a node's IP address. It uses a word w as a key and each FCache entry is a data structure that holds an IP address. Thus, whenever P needs to contact another node P' that is responsible for queries containing w, it searches its FCache. If FCache contains an entry for w, P can directly contact P' using the IP stored in its FCache. If w is not contained in FCache, P uses the standard DHT lookup protocol to locate P' and stores contact information in FCache for further reference. Using FCache, the cost of processing a published resource p is reduced to $O(v + (h - v) \log N)$, where v is the number of words of p contained in FCache. Notice that the construction and maintenance of FCache comes at no extra message cost and node routing information is discovered only when needed. In the experiments presented in the next section we discuss good choices for FCache size (see Sect. 4.4).

The only extra cost involved with FCache is due to possible cache misses because of network dynamicity. In an FCache miss, the node needs to utilise the routing infrastructure at the cost of $O(logN)$ messages to locate a node. However, the new contact information is used to update the FCache entry for future reference. Misses are most likely to occur for infrequent words, since nodes responsible for storing queries with frequent words will be contacted repeatedly.

3.6   Network Dynamicity and Fault Tolerance

The issues introduced by the dynamic nature of P2P systems may be distinguished in two general categories: (i) topology changes as nodes move in and out of the system and (ii) content changes as users shift their interests to new topics while losing interest in others.

In a dynamic network, nodes may join, leave, or fail at any time (referred to as *node churn* in the literature). The main challenge in dealing with these situations in a DHT is preserving the ability to locate every key in the network. The stabilisation protocol provided by Chord aggressively maintains the finger tables of all nodes as the network evolves, by relying on successor pointers to ?undertake correctness of lookups and finger table repairs. This stabilisation scheme guarantees to offer reachability of existing nodes even at the face of concurrent joins, leaves, or fails and allows lookups to be both fast and correct. Since all nodes are uniquely identified in the network, and the Chord identifier calculated is the same for each reconnection, a node is naturally mapped at the same location on the Chord ring every time. This is exploited by the DHTrie protocols to store notifications for a node at its successor and to deliver them upon node reconnection. Naturally, successor nodes are also used for data handover when a node departs normally from the network. To cope with data loss due to node failures and accelerate lookups further, replication [60–62] and caching [63, 64] algorithms may be utilised. Finally, note that changes in network topology will also lead to FCache misses (remember that misses do not affect the correctness of the protocols) and hence, increase message traffic, as shown in Sect. 4.8.

Naturally, the interests of the nodes evolve over time resulting in creating, modifying, or deleting queries from the network, or even changing the topic and rate of their publications. These changes will cause an increase in message traffic as long as the network tries to cope with the content shift. Newly introduced topics or topics that have suddenly gained increasing interest will introduce new terms, which in turn will be infrequent at the beginning, but their frequency of occurrence will increase with user publications. These changes in content are expected to initially generate FCache misses (i.e., increase network traffic), but as specific terms become popular FCache will be gradually updated.

Table 1. Some key characteristics of the two corpora used for the evaluation

Description	NN corpus	DBP corpus
Collection size compressed (uncompressed)	99.6 MB (346.2 GB)	548.8 MB (2 GB)
Number of documents	10,426	3,144,265
Document vocabulary size in words	379,484	2,902,491
Maximum document size in words (KB)	104,500 (595.5 KB)	15,815 (150.3 KB)
Minimum document size in words (KB)	26 (0.2 KB)	1 (0.002 KB)
Average document size in words (KB)	5,415 (32.9 KB)	91 (0.5 KB)

4   Experimental Evaluation

To carry out the experimental evaluation of the protocols described in the previous section, we needed metadata for incoming resources, as well as user queries. For the model $\mathcal{AWPS}$ considered in this work there are various document sources that one could consider: TREC corpora, metadata for papers on various publisher Web sites (e.g., ACM or IEEE), electronic newspaper articles, articles from news alerts on the Web (http://www.cnn.com/EMAIL), and others. However, it is rather difficult to find user queries except by obtaining proprietary data (e.g., from CNN's news or Springer's journal alert system). Additionally, notice that using query logs of one-time queries as continuous queries does not create realistic query databases. One-time queries are in general short and focused, as they express one-time information needs, while continuous queries tend to be longer, more complex and more general, in order to satisfy long-term information needs.

4.1   Experimental Setup

In this section, we describe the document and query sets used to evaluate our methods, and present the performance criteria and setup of our evaluation.

Document Corpora. For our experiments, we used two sets of real-life documents and queries. The first set is composed of 10,426 documents downloaded from CiteSeer (http://citeseer.ist.psu.edu), originally compiled in [65], and used also in [10, 52, 57]. These documents are research papers in the area of Neural Networks; we will refer to them as the *NN corpus*. To assess the generality of our approach, we have also conducted experiments with a larger and more varied corpus. The dbpedia (http://dbpedia.org) corpus –we will refer to it as *DBP corpus*– consists of more than 3 million documents that are extended abstracts from the Wikipedia website. The DBP corpus was chosen due to its differences to the NN corpus (smaller average document size, larger diversity in topics, wider vocabulary) and is used to demonstrate the performance of our protocols under a different setting. Table 1 summarises some key characteristics of the two corpora used in the evaluation.

All the experiments shown in this section were carried out using both document corpora. However, due to space considerations we report graphs for both corpora only when there exists a notable difference between the two experiments.

Query Sets. Since no database of continuous queries was available to us, we used two different methodologies to create continuous queries under model $\mathcal{AWPS}$.

The queries for the NN corpus are synthetically generated and consist of two parts: (i) a Boolean part containing atomic Boolean queries of the form $A \sqsupseteq wp$ and (ii) a VSM part containing atomic queries of the form $A \sim_k s$, where s is a text value. We set A to be TITLE, AUTHORS, ABSTRACT, or BODY with some probability. Subsequently, each atomic Boolean query of the form $A \sqsupseteq wp$ is generated using *words* and *technical terms* extracted automatically from the NN corpus using the C-value/NC-value approach of [66]. For more details of the methodology the interested reader can refer to [10,52,57]. An example of a user query created synthetically from the methodology briefly sketched above is:

$$(\text{AUTHOR} \sqsupseteq Darwen) \quad \wedge$$
$$(\text{TITLE} \sqsupseteq implementation \wedge (RBF \prec_{[0,3]} networks)) \quad \wedge$$

ABSTRACT $\sim_{0.6}$ "Most work on the evolutionary approach to the iterated ... "

Since there is no publicly available database of continuous queries for the DBP corpus, we used 20.2 million Wikipedia article titles and categories (modified appropriately to fit our query language) as user queries. Each title or category represents one continuous query q that contains either a Boolean or a VSM part. For the case of the DBP corpus, we avoided synthetic creation of more complex queries (as done before for the case of the NN corpus) in order to demonstrate the performance of our methods under a different query setting.

Setup. We have implemented and experimented with eight variations of the DHTrie protocols: the iterative method It, the recursive method Re, the hybrid method $fixHy$, and the continuous splitting method Spl, which do not employ an FCache, and their counterparts that utilise an FCache (ItC, ReC, $fixHyC$, and $SplC$ respectively). The experiments with both corpora were conducted using the same machinery to enable the comparison across the different settings. All the methods and the DHTrie simulator were implemented in C/C++.

To carry out each experiment described in this section, we execute the following steps. Initially the network is set up by assigning keys to nodes. These keys are calculated using the SHA-1 cryptographic hash function and randomly created IP addresses and ports. After the network set up, we create 5M user queries and distribute them among the nodes using the protocol described in Sect. 3.1. According to the publication protocol, the number of posted queries does not affect the cost for publishing a document in the network; it only affects the matching time for the local filtering algorithms and the number of matching notifications produced (i.e., the higher the number of posted queries is, the higher the number of matching notifications produced). Table 2 summarises the parameters and the baseline values used for the experiments.

Evaluation Metrics. We are mainly interested in the performance of the eight different protocols in terms of *network traffic* and *latency* to publish a document or subscribe a query. To measure network traffic, we publish the corpus documents at different nodes and record the network activity. In our network, we can distinguish between two types of messages: messages sent through the DHT infrastructure and messages sent to a node using directly its IP address (FCache

Table 2. Parameters varied in experiments, their descriptions, and their baseline values

Parameter	Description	Baseline value
N	# of nodes in the system	10 K–100 K
Q	# of queries assigned to nodes	5 M
C_s	# of entries in FCache	30 K
C_t	# of publications used to train FCache	10 K
W	Average # of words per published document	5415 (NN), 91 (DBP)
SF	Split factor (used for load balancing)	1, 10, 20, 30
T	Split threshold (used for load balancing)	10
σ	Size of recipients list ($fixHy$ method)	50 (NN), 10 (DBP))
π	Percentage of recipients list ($perHy$ method)	4 %

messages). In our experiments, we record and present the effects of both types of messages. Latency is measured in number of hops as follows. For each message (either publication or subscription) initiated by node P, we record the longest chain of messages needed until the message reaches all the intended recipients.

4.2   Varying the Type of Queries

The first set of experiments investigates the cost of indexing a query in the network. For this setup we used two types of queries: (i) queries including only vector space atomic parts and (ii) queries with both Boolean and vector space parts. Indexing the second type of queries is the same as indexing queries with Boolean atomic parts only (see Sect. 3.1).

Each bar in Fig. 3 shows the average message traffic recorded when indexing 500 K queries of each type in a network of 50 K nodes for both corpora. The most important observation in these graphs is that, regardless of the protocol and corpus applied on, vector space queries are more expensive to index than Boolean or mixed type queries. This happens because vector space queries are indexed at all nodes responsible for the distinct words in the query, contrary to other query types that are indexed under only one node (see Sect. 3.1). Notice the important role of FCache, the use of which manages the forwarding to the intended recipients of more than 1/3 of the total network traffic, thus relieving the DHT infrastructure of substantial messaging effort. It is clear that ReC and $fixHyC$ are the best performing protocols for vector space query indexing in terms of message traffic. The difference in the message traffic induced by the queries of the two corpora is attributed to the different query lengths.

Figure 4 presents the publication latency achieved by all our protocols when indexing 500 K (vector space or mixed) queries in a network of 50 K or 100 K nodes for both corpora. As we can see in this figure, latency in the indexing of Boolean or mixed type queries is invariant since they are indexed under only one node. For the vector space queries however, one important observation is the low latency of the iterative and the continuous splitting methods, and the high latency of the recursive ones. This is due to the routing infrastructure used and the specifics of each method. The iterative methods use a single lookup message for each one of the intended recipients of the query, thus parallelising the subscription process. The continuous splitting methods split the recipient lists

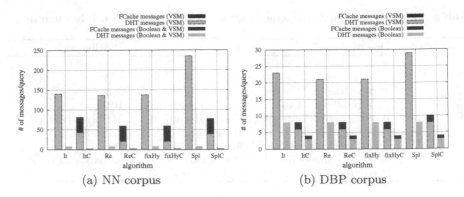

Fig. 3. Message traffic for indexing a query in the network

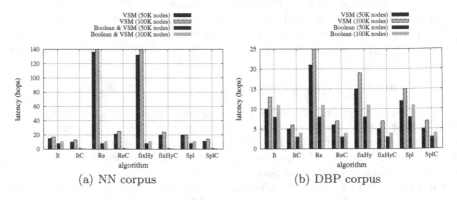

Fig. 4. Latency for indexing a query in the network

and adapt the subscription process to the finger tables of the forwarding nodes. On the other hand, the recursive method uses long recipients lists and contacts them in a recursive way, thus increasing subscription latency. Additionally, protocol $fixHy$ seems to behave similarly to Re in terms of latency, which is explained by the fact that in this experiment the two protocols have roughly the same size of recipients lists. This happens because Re creates small recipients lists (due to the size of the query) and thus, the size of the list is similar to the size we use for protocol $fixHy$. FCache, similarly to message traffic, plays an important role by reducing latency (up to 60 %) for all the protocols.

4.3 Varying Network Size

Although query indexing performance is important, in an IF scenario resource publication is the time critical component. This second set of experiments targets the performance of the protocols in terms of message traffic and publication latency for different network sizes. In this experiment, we randomly selected 100 documents from the NN and the DBP corpus and used them as publications by randomly assigning each one to a different publisher node for each of the 10 dif-

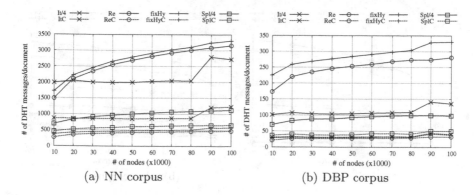

(a) NN corpus   (b) DBP corpus

Fig. 5. Message traffic for various network sizes

ferent runs used for averaging measurements. Having published the documents, we recorded the total number of DHTrie messages generated by the network in order to match these documents against the indexed user queries.

In Fig. 5, the performance of the protocols in terms of DHTrie messages/ document for both corpora is shown. The main observation is that the number of messages generated by all protocols grows at a logarithmic scale mainly due to the routing infrastructure used. A second observation emerging from the graph is the effectiveness of the FCache independently of the message routing protocol used and the corpus it is applied at. The use of FCache results in the reduction of messages sent using the routing infrastructure by more than 6 times for NN corpus (resp. 7 times for the DBP corpus) in the recursive, the hybrid and the continuous splitting method, and by 8 times for the NN corpus (resp. 4 times for the DBP) in the iterative method. Notice that the improvement in the performance of the protocols when using the FCache is slightly lower for the DBP corpus (compared to the NN corpus) due to the significantly smaller document size and the wider vocabulary (because of lower FCache utilisation and thus higher DHT traffic). Finally, notice that the number of DHT messages needed to index a document from the DBP corpus is significantly lower than that of the NN corpus, due to the significantly smaller average document size.

In Fig. 6, the performance of the different protocols in terms of publication latency for both corpora is shown. Similarly to the previous set of experiments, low latency is observed when using the iterative or the continuous splitting methods, whereas high latency is caused by the recursive ones. The use of the FCache reduces publication latency for both corpora by shortening the intended recipients lists of ReC and $fixHyC$. Additionally, it is worth pointing out that the smaller document size of the DBP corpus results in lower publication latency compared to that of the NN corpus, due to the smaller recipients lists.

Finally, in our measurements (graph not shown due to space reasons) the lowest processing cost per document for a network size of 100 K nodes for the NN corpus (resp. DBP corpus) is presented for method ReC with about 1,300 (resp. 42) messages in total, with about 65 % of them being FCache messages, as opposed to 40 % for method $SplC$, and 55 % for method ItC for both corpora.

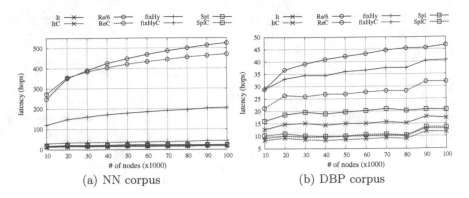

(a) NN corpus

(b) DBP corpus

Fig. 6. Latency for various network sizes

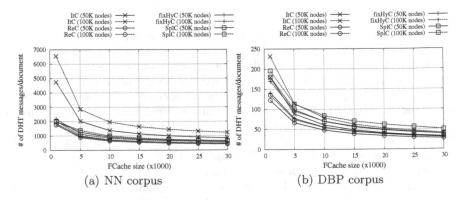

(a) NN corpus

(b) DBP corpus

Fig. 7. Message traffic for different FCache sizes

4.4   Varying the FCache Size

The third set of experiments targeted the performance of the protocols under different FCache sizes, and studied the effect of FCache in message traffic and publication latency. Initially, we used (a part of) the document corpora as training sets for populating the FCache of the different nodes; a randomly chosen node P publishes 10 K documents and populates its FCache with the IP addresses of the nodes that are responsible for the most frequent words contained in the published documents. Then, another 100 documents are published by P and the size of the FCache is limited to different values. Subsequently, the total number of messages used to match these documents against the stored user queries is recorded and averaged over 10 runs with different nodes. Figure 7 shows the messages traffic per document for the two corpora as the size of the FCache grows.

As shown in Fig. 7, the number of messages sent using the DHTrie routing infrastructure reduces quickly for both corpora as the size of FCache increases, and the decrease rate depends on FCache size due to the skewness in the corpus vocabulary. This results in reaching an FCache size after which no significant

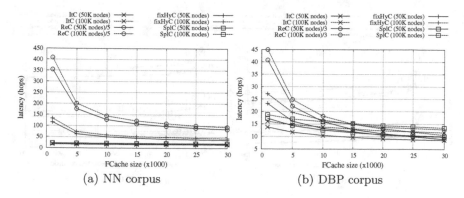

Fig. 8. Latency for different FCache sizes

effect is observed in message traffic reduction (around 30 K entries, the rightmost point on the x-axis). Additionally, the reduction factor for all methods and for both corpora is similar: 40–50 % (resp. 10–15 %) reduction in message costs depending on the method for small (resp. large) FCache sizes. Notice also that for protocols *ReC*, *HyC*, and *SplC* the performance of FCache remains almost constant for different network sizes, whereas for protocol *ItC* 50 % more DHTrie messages/document are needed for an 100 % increase in network size. Finally, notice that the number of DHT messages needed to index a document from the DBP corpus is significantly lower than that of the NN corpus due to the significant difference in average document size.

In Fig. 8, we show the publication latency for the different protocols and the way it is affected by the variation of the FCache size. As expected, the reduction in the latency for all the protocols is lower as the FCache size increases due to the skewness of the vocabulary entries used for populating the FCache. Additionally, not all protocols are affected in the same way from FCache increase in size. *ItC* and *SplC* remain relatively unaffected for both corpora by the increase both in FCache size and in network size, something that is also verified from the graphs of the previous section. This is due to the routing infrastructure and the parallel way of publishing the incoming documents. Contrary, protocols *ReC* and *fixHyC* seem to perform better when the size of the FCache increases, since this causes reduction in the size of recipients lists. Moreover, the reduction factor across corpora is similar: low for methods *ItC* and *SplC*, while it reaches 35–45 % (resp. 5–15 %) for small (resp. large) FCache sizes for methods *ReC* and *fixHyC*. Finally, FCache is equally utilised by all methods (graph omitted due to space reasons), as the number of FCache messages/document is similar for all methods and reaches up to 900 (resp. 19) FCache messages/document for the NN corpus (resp. DBP corpus) for 30 K entries.

4.5 Effect of FCache Training

In this set of experiments, we measure the effect of FCache training on message cost and publication latency in the NN corpus (results for the DBP corpus are similar and are omitted due to space reasons). To do so, we randomly selected

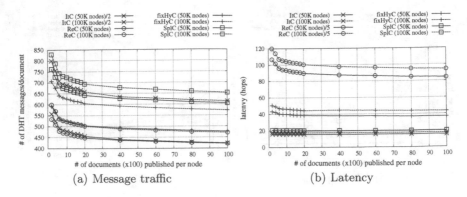

(a) Message traffic   (b) Latency

Fig. 9. Performance for different levels of FCache training

a node P and trained its FCache with a varying number of documents. In this way, the node was able to collect statistics about frequent words used in document publications and populate its FCache with pointers to frequently contacted nodes. Subsequently, we published 100 documents to P and recorded the average message cost and publication latency. The results shown in Fig. 9 are averaged over 100 runs for different nodes to eliminate network topology effects.

Figure 9(a) shows that the performance of all protocols improves as more documents get published. Methods ReC and $fixHyC$ are less sensitive in this parameter, as the difference in the number of messages observed is about 100 messages for 50 times more documents (the leftmost and rightmost point in the x-axis). Additionally, ReC and $SplC$ show less sensitivity with respect to the network size, contrary to ItC that needs about 50 % more messages. Finally, all methods show a similar behaviour for the two network sizes tested.

Figure 9(b) shows the effect of the number of publications in latency. We observe that method ReC is the most affected by the training level of the FCache, as it is heavily dependent on the FCache information to reduce long recipients lists. Method $fixHyC$ is less affected as it produces shorter recipient lists than ReC, while methods $SplC$ and ItC remain unaffected due to the protocol design. Additionally, all methods present a slight increase in message traffic when doubling the network size due to the logarithmic routing.

Finally, the number of FCache hits for the NN corpus (resp. DBP corpus) and for all methods is between 830 and 875 (resp. 12 and 20) messages/document for a network of 50 K nodes. This shows that FCache hits are only affected by the size and skewness of the published data, not by the protocol used.

4.6   Varying the Document Size

Document (i.e., publication) size is an important parameter in the performance of our protocols. This set of experiments targeted the performance of the protocols under various document sizes. Due to space reasons we show only the experiments on the NN corpus that has a larger variation in document size. The findings for the DBP corpus are briefly summarised in the text and are in line with the ones presented here. Figure 10 shows the message cost and latency for

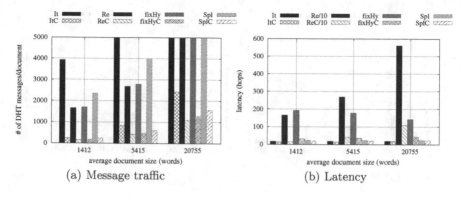

(a) Message traffic   (b) Latency

Fig. 10. Performance for documents of different size

publishing documents of varying size for all protocols. Each bar is an average of the message cost and latency (appropriately truncated to show the best performing methods) for 100 documents, published by 1,000 different nodes (in a network of 50 K nodes in total) to normalise network topology effects.

Figure 10(a) shows that for small documents, methods Re, $fixHy$, and Spl achieve 50 % less message traffic than It, while all FCache variations of the protocols perform similarly. This happens because in smaller documents there will be less infrequent words that may result in FCache misses. However, as document size increases the importance of the message forwarding method is more obvious (i.e., notice that ReC is able to process documents of 21 K words by using only 1,000 messages). Note also that although protocols ReC, $fixHyC$, and $SplC$ perform similarly in terms of message traffic, as discussed later in this section, $SplC$ handles latency better than its counterparts. Our findings for the DBP corpus are similar, but, since the average document size is significantly smaller, message traffic is about 10 times lower (see also Fig. 5).

Figure 10(b) shows how document size affects latency for the different protocols. The most important observation is the inefficient performance of the Re and ReC protocols (notice that measurements are reduced by a factor of 10 for readability), which shows the dependence of both methods on document size (that increases the size of recipients lists). Contrary, the rest of the protocols are insensitive to document size, for different reasons each: It and ItC because of the lack of recipient lists, $fixHy$ and $fixHyC$ because of document size-independent recipient lists, and Spc and $SplC$ because of the adaptivity of the forwarding process. The measurements for the DBP corpus showed a similar behaviour for all methods due to the small average document size and thus recipient list size.

We also examined the relative increase in message traffic and latency for three groups of documents, D_1, D_2, and D_3, where D_2 is 3 times larger and D_3 is 14 times larger on average than D_1. Initially, 100 random nodes were chosen to publish documents from group D_1, and the message traffic and latency were recorded. Then, the other two document groups were published in the same way, and the measurements were recorded and compared to those of group D_1. Figure 11 shows the factor of increase in message traffic and latency for each protocol when publishing the two different groups of documents. Our findings on

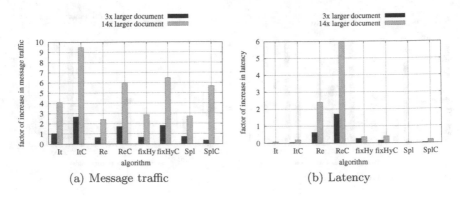

Fig. 11. Increase rate for different document sizes

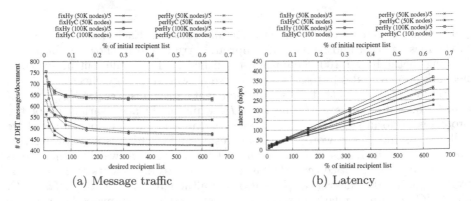

Fig. 12. Performance when varying σ ($fixHy$) and π ($perHy$)

the sensitivity of the methods to document size are aligned with those of Fig. 10: in terms of message traffic ReC, $fixHyC$, and $SplC$ show low sensitivity to document size, while in terms of latency Re and ReC are highly sensitive.

4.7   Comparison of the Hybrid Methods

This set of experiments aims at comparing message overhead and publication latency of the hybrid method variants by examining the correlations between the parameters σ and π of methods $fixHy$ and $perHy$, and the performance of the parameterless method $medHy$. Figure 12 demonstrates the performance of the $fixHy$ and $perHy$ methods for two different network sizes (50 K and 100 K nodes). Each point is averaged over 10 runs, and 100 NN corpus documents, randomly assigned to publisher nodes, were used as incoming publications. The findings for the DBP corpus are similar and are omitted for space reasons.

Figure 12(a) shows the average number of DHT messages needed to publish a large document as the desired recipient list size increases for the hybrid methods. To interpret the results of this graph the reader is reminded that the hybrid protocols try to combine the iterative and recursive protocols: the shorter

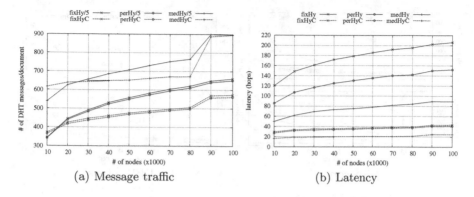

Fig. 13. Performance of the hybrid methods for various network sizes

the recipients list size is, the closer the protocol is to the iterative counterpart (notice network traffic reduction as the recipient list size increases). Additionally, FCache reduces network traffic and the effect of network size significantly, while making the methods less sensitive to parameter changes. Finally, notice that although the splitting of the recipients list is performed in a different way by $fixHy$ and $perHy$, parameters σ and π have a similar effect since they relate through the average publication size (i.e., it is possible to adjust π so that in the average case it will split the message in pieces of average size σ). Notice however, that π is easier to set than σ as it does not require any knowledge on the specifics of the published documents.

Publication latency is linear to the increase in the recipient list size (Fig. 12(b)) for both methods, while FCache manages to keep latency low.

Finally in Fig. 13(a) and (b), we demonstrate message traffic and latency for all hybrid variations. Message traffic for all methods grows logarithmically due to the routing infrastructure, while the introduction of FCache results in a significant decrement in message traffic. As the $medHy$ method is by design aimed towards optimising latency (due to the way of splitting recipients lists), $fixHyC$ and $perHyC$ have also been set with latency in mind ($\sigma = 50$, $\pi = 0.04$) for comparison reasons. Finally, notice that the $fixHy$ and $perHy$ methods perform similarly in terms of message traffic, but differ in latency, which demonstrates the importance of optimising this tradeoff and constituted the main driver for the introduction of Spl and $SplC$ methods.

4.8   Effect of Node Churn

In this section, we target the performance of the protocols in terms of message traffic and publication latency under node churn by introducing a short life span for a varying percentage of nodes in the network.

In Fig. 14, our measurements show that when 5 % of the nodes are off-line during a lookup, the message cost increase is no more than 8 % for the NN corpus (12 % for the DBP corpus), showing that FCache is able to cope up with misses. On the other hand, when 30 % of the nodes are off-line, the message cost increases significantly for both corpora, since for each FCache miss several DHT messages

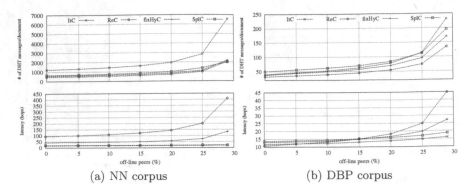

(a) NN corpus   (b) DBP corpus

Fig. 14. Message traffic and latency for 100 K nodes under churn

have to be issued. Remember though, that FCache misses affect only network traffic and not the correctness of the protocols. Additionally, all methods, apart from *ReC* (that packs messages together and increases the recipients list size), present a good overall performance in terms of latency.

To deal with node failures for methods that do not rely on FCache (i.e., *It*, *Re*, *fixHy*, and *Spl*) we reside on DHT mechanisms, which can guarantee correctness of lookups and finger table repairs. All measurements for node churn in Chord [56] carry over to our setting; for more details the reader is referred to [56] due to space reasons.

4.9   Skewed Data Distributions and Load Balancing

In typical IR scenarios the word distributions associated with documents and queries are typically skewed. In a pub/sub setting, *load balancing* becomes a key issue when trying to partition the query space among the different nodes of a DHT. We can distinguish three types of node load: *query load* (i.e., the number of queries stored at a node), *routing load* (i.e., the number of messages a node forwards due to the protocols), and *filtering load* (i.e., the number of publications a node has to filter against the stored queries).

Balancing the Filtering Load. In the DHT literature, work on load balancing has concentrated on two particular problems: (i) *address-space load balancing* concerning how to partition the address-space of a DHT "evenly" among keys; it is typically solved by relying on consistent hashing and constructions such as virtual servers [8] or potential nodes [67] and (ii) *item load balancing* addressing how to balance load in the presence of data items with arbitrary load distributions [67,68] as in our case.

We have implemented and evaluated a simple algorithm based on the well-known concept of *load-shedding* (LS), where an overloaded node attempts to off-load work to less loaded nodes. Once a node P understands that it has become overloaded, it chooses the most frequent word w it is responsible for and a small integer k. Then P contacts the nodes responsible for words wj for all j, $1 \leq j \leq k$, where wj is the concatenation of strings w and j, and asks

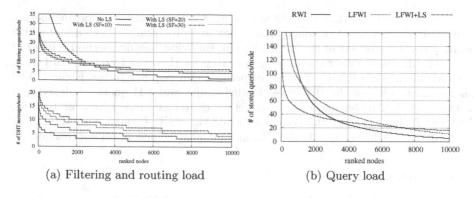

(a) Filtering and routing load   (b) Query load

Fig. 15. Load distribution for the 10 K most loaded nodes

them to be its replicas. Then P notifies the rest of the network about this change in responsibilities by piggy-backing the necessary information in DHTrie maintenance messages. Each node M that receives this message notes down the word w. Later on, if M has a new publication containing w, it divides the filtering responsibility for w among P and k other nodes by concatenating a random number from 1 to k to the end of w and using DHTrie to find the node responsible for the concatenated word. In this way, the filtering responsibility of w for P is reduced by $k+1$ times (node P and k new nodes). We call $k+1$ the *split factor* (*SF*) in subsequent experiments.

Figure 15(a) shows the average number of filtering requests (top part) received by each node in a time window T for a period of 100*T. *SF* was varied between 10 and 30 nodes and T was set to 10 filtering requests. We also varied T but did not observe significant differences in the load distribution.

Balancing the Routing Load. Balancing the filtering load causes an increase in message traffic due to FCache misses and thus, the overall routing load of the system is increased. Figure 15(a) shows the number of routing requests (bottom part) received by the 10 K most loaded nodes in the network. The number of DHT messages/document increases after the load balancing algorithm is run (80 % for SF=10, 180 % for SF=20, and 240 % for SF=30), but the new load imposed on the network is well distributed among the nodes and does not cause overloading in any specific group of nodes.

Balancing the Query Load. Even query distribution among nodes is a hard task to achieve since typically queries follow a skewed word distribution. To distribute the queries to the nodes responsible, we utilised two different query indexing methods. The first method, coined *RWI* (Random Word Index), follows the subscription protocol of Sect. 3.1, where a node P indexes a query q to the node responsible for a *randomly* selected word w contained in any of the text values $s_1, \ldots, s_m$ or word patterns $wp_{m+1}, \ldots, wp_n$ of q. Contrary, the second method, coined *LFWI* (Least Frequent Word Index), takes into account the *document frequency* of the words contained in q and indexes q to the node responsible for the *least frequent* word w contained in it.

Notice that the methods described above are orthogonal to the routing method utilised by P to forward the query, since they are only used to select under which word q will be indexed in the network. The intuition behind method $LFWI$ is to index the query under the node responsible for the least frequent word in it, thus avoiding the overload of nodes responsible for popular terms. Figure 15(b) shows the results for the 10 K most loaded nodes and 1 M queries indexed in a network of 50 K nodes and each graph is produced as an average over 10 runs. Method $LFWI+LS$, i.e., the combination of the $LFWI$ and LS methods (for with SF=10), achieves the most uniform load distribution of all approaches.

4.10   Summing Up

In all experiments, the methods with the FCache (ItC, ReC, $fixHyC$, $SplC$) outperformed (both in message traffic/latency) their counterparts without the FCache (It, Re, $fixHy$, Spl) showing the usefulness of the proxying mechanism.

When message traffic is the optimisation metric (at the expense of latency), method ReC is the best candidate, as it is less sensitive to network and publication size. Contrary, when latency is the optimisation metric (at the expense of message traffic), method ItC presents the best alternative, as it is less affected by network size, FCache size/training, and publication size. Hybrid methods $fixHyC$ and $perHyC$ are tunable alternatives to the ItC and ReC methods, adjustable to publication size, and offer a good tradeoff between message traffic and latency, while method $medHyC$ is the parameterless variation of the hybrid family that slightly favors latency over message traffic.Finally, method $SplC$ is slightly more expensive than ReC in network traffic, but its latency is as low as the best performing method ItC. Moreover, $SplC$ is less sensitive than the hybrid methods to changes in network size and FCache size/training.

Overall, $perHyC$ and $SplC$ are the two most versatile and well-performing protocols that put emphasis both on optimising message traffic and latency. Method $perHyC$ may be adjusted in a per-node fashion, however parameter setting may require background knowledge of publication characteristics. On the other hand, $SplC$ is an adaptable and versatile method that performs well under many different scenarios (including node churn) and can be deployed off-the-shelf, without any need for parameter setting.

5   Conclusions and Outlook

In this work, we have presented and evaluated a set of protocols that efficiently extend Chord with pub/sub functionality and introduced proxying and load balancing mechanisms to cope with message traffic, latency, and skewness of data. The results of the earlier version of this paper [10] have influenced most of our work on P2P computing over the last years, inspiring us to develop IF functionality [27] in the Minerva system [53], study IF in an XML context [69], design DHT-based digital libraries [52], and implement Web/Grid service registries [70] for the EU projects OntoGrid and SemsorGrid4Env.The deployment of these ideas on various domains demonstrates the generality of the problem and shows that our protocols may be applied beyond the adopted scenario.

Lately, MapReduce [71] is widely used as the programming paradigm to achieve distributed data analysis, load balancing, and fault tolerance by parallelising map and reduce operations in the cloud.We plan to port our work to the MapReduce paradigm (e.g., following the philosophy of Memcached for a generic distributed service) to allow the deployment of our protocols in a well-known computing paradigm aiming for higher penetration in domains other than P2P. Additionally, such an implementation will encourage the usage and evaluation of the protocols in real-life scenarios and allow us to get usage data involving performance measurements, real user profiles, and publishing behaviour patterns.

Furthermore, the deployment of our protocols in large-scale distributed social networks would allow novel data management functionality, like subscriptions over content/tags with aggregation (e.g., notify me when a published document matches my continuous query and k of my friends have tagged it as interesting).

References

1. Hameurlain, A., Hussain, F.K., Morvan, F., Tjoa, A.M. (eds.): Globe 2012, vol. 7450. Springer, Heidelberg (2012)
2. Sinha, V., Gupta, A., Kohli, G.S.: Comparative study of P2P and cloud computing paradigm usage in research purposes. In: Das, V.V., Stephen, J., Chaba, Y. (eds.) CNC 2011. CCIS, vol. 142, pp. 341–347. Springer, Heidelberg (2011)
3. Kavalionak, H., Montresor, A.: P2P and cloud: a marriage of convenience for replica management. In: Kuipers, F.A., Heegaard, P.E. (eds.) IWSOS 2012. LNCS, vol. 7166, pp. 60–71. Springer, Heidelberg (2012)
4. Trajkovska, I., Salvachua Rodriguez, J., Mozo Velasco, A.: A novel P2P and cloud computing hybrid architecture for multimedia streaming with QoS cost functions. In: ACM Multimedia (2010)
5. Kontominas, D., Raftopoulou, P., Tryfonopoulos, C., Petrakis, E.G.: DS4: a distributed social and semantic search system. In: ECIR (2013)
6. Loupasakis, A., Ntarmos, N., Triantafillou, P.: eXO: decentralized autonomous scalable social networking. In: CIDR (2011)
7. Graffi, K., Gross, C., Mukherjee, P., Kovacevic, A., Steinmetz, R.: LifeSocial.KOM: a P2P-based platform for secure online social networks. In: P2P (2010)
8. Stoica, I., Morris, R., Karger, D., Kaashoek, M., Balakrishnan, H.: Chord: a scalable peer-to-peer lookup service for internet applications. In: ACM SIGCOMM (2001)
9. Koubarakis, M., Skiadopoulos, S., Tryfonopoulos, C.: Logic and computational complexity for boolean information retrieval. IEEE TKDE 18(12), 1659–1666 (2006)
10. Tryfonopoulos, C., Idreos, S., Koubarakis, M.: Publish/Subscribe functionality in IR environments using structured overlay networks. In: ACM SIGIR (2005)
11. Carzaniga, A., Rosenblum, D.S., Wolf, A.: Design and evaluation of a wide-area event notification service. ACM TOCS 19(3), 332–383 (2001)
12. Koubarakis, M., Tryfonopoulos, C., Idreos, S., Drougas, Y.: Selective information dissemination in P2P networks: problems and solutions. SIGMOD Rec. 32(3), 71–76 (2003)
13. Rowstron, A., Kermarrec, A.-M., Druschel, P.: SCRIBE: the design of a large-scale event notification infrastructure. In: Crowcroft, J., Hofmann, M. (eds.) NGC 2001. LNCS, vol. 2233, pp. 30–43. Springer, Heidelberg (2001)
14. Pietzuch, P., Bacon, J.: Hermes: a distributed event-based middleware architecture. In: DEBS (2002)

15. Tam, D., Azimi, R., Jacobsen, H.-A.: Building content-based publish/subscribe systems with distributed hash tables. In: Aberer, K., Koubarakis, M., Kalogeraki, V. (eds.) DBISP2P 2003. LNCS, vol. 2944, pp. 138–152. Springer, Heidelberg (2004)
16. Terpstra, W., Behnel, S., Fiege, L., Zeidler, A., Buchmann, A.: A peer-to-peer approach to content-based publish/subscribe. In: DEBS (2003)
17. Gedik, B., Liu, L.: PeerCQ: a decentralized and self-configuring peer-to-peer information monitoring system. In: ICDCS (2003)
18. Karger, D., Lehman, E., Leighton, T., Levine, M., Lewin, D., Panigrahy, R.: Consistent hashing and random trees: distributed caching protocols for relieving hot spots on the World Wide Web. In: ACM STOC (1997)
19. Bender, M., Bender, M., Michel, S., Michel, S., Parkitny, S., Parkitny, S., Weikum, G., Weikum, G.: A comparative study of pub/sub methods in structured P2P networks. In: Moro, G. (ed.) DBISP2P 2005 and DBISP2P 2006. LNCS, vol. 4125, pp. 385–396. Springer, Heidelberg (2007)
20. Triantafillou, P., Aekaterinidis, I.: Content-based publish-subscribe over structured P2P networks. In: DEBS (2004)
21. Gupta, A., Sahin, O.D., Agrawal, D.P., El Abbadi, A.: Meghdoot: content-based publish/subscribe over P2P networks. In: Jacobsen, H.-A. (ed.) Middleware 2004. LNCS, vol. 3231, pp. 254–273. Springer, Heidelberg (2004)
22. Aekaterinidis, I., Triantafillou, P.: PastryStrings: a comprehensive content-based publish/subscribe DHT network. In: ICDCS (2006)
23. Aekaterinidis, I., Triantafillou, P.: Internet scale string attribute publish/subscribe data networks. In: CIKM (2005)
24. Tran, D., Pham, C.: Enabling content-based publish/subscribe services in cooperative P2P networks. Comput. Netw. 54(11), 1739–1749 (2010)
25. Lo, S.C., Chiu, Y.T.: Design of content-based publish/subscribe systems over structured overlay networks. IEICE Trans. E91–D(5), 1504–1511 (2008)
26. Liau, C.Y., Ng, W.S., Shu, Y., Tan, K.-L., Bressan, S.: Efficient range queries and fast lookup services for scalable P2P networks. In: Ng, W.S., Ooi, B.-C., Ouksel, A.M., Sartori, C. (eds.) DBISP2P 2004. LNCS, vol. 3367, pp. 93–106. Springer, Heidelberg (2005)
27. Tryfonopoulos, C., Zimmer, C., Koubarakis, M., Weikum, G.: Architectural alternatives for information filtering in structured overlay networks. IEEE Internet Comput. 11(4), 24–34 (2007)
28. Zheng, X., Luo, J., Cao, J.: Pat: a P2P based publish/subscribe system for QoS information dissemination of web services. In: ICWS (2009)
29. Cheung, A.Y., Jacobsen, H.A.: Load balancing content-based publish/subscribe systems. ACM TOCS 28(4), 46–100 (2010)
30. Bernard, S., Potop-Butucaru, M.G., Tixeuil, S.: A framework for secure and private P2P publish/subscribe. In: Dolev, S., Cobb, J., Fischer, M., Yung, M. (eds.) SSS 2010. LNCS, vol. 6366, pp. 531–545. Springer, Heidelberg (2010)
31. Drosou, M., Stefanidis, K., Pitoura, E.: Preference-aware publish/subscribe delivery with diversity. In: DEBS (2009)
32. Tang, C., Xu, Z.: pFilter: global information filtering and dissemination using structured overlays. In: FTDCS (2003)
33. Zhu, Y., Hu, Y.: Ferry: a P2P-based architecture for content-based publish/subscribe services. IEEE TPDS 18(5), 672–685 (2007)
34. Ratnasamy, S., Francis, P., Handley, M., Karp, R., Shenker, S.: A scalable content-addressable network. In: ACM SIGCOMM (2001)
35. Kossmann, D.: The state of the art in distributed query processing. ACM Comput. Surv. 32(4), 422–469 (2000)

36. Stonebraker, M., Aoki, P., Litwin, W., Pfeffer, A., Sah, A., Sidell, J., Staelin, C., Yu, A.: Mariposa: a wide-area distributed database system. VLDB J. **5**(1), 48–63 (1996)
37. Litwin, W., Neimat, M.A., Schneider, D.A.: LH* - a scalable, distributed data structure. ACM TODS **21**(4), 480–525 (1996)
38. Balakrishnan, H., Kaashoek, M., Karger, D., Morris, R., Stoica, I.: Looking up data in P2P systems. CACM **46**(2), 43–48 (2003)
39. Huebsch, R., Hellerstein, J., Lanham, N., Loo, B., Shenker, S., Stoica, I.: Querying the internet with PIER. In: VLDB (2003)
40. Harren, M., Hellerstein, J.M., Huebsch, R., Loo, B.T., Shenker, S., Stoica, I.: Complex queries in DHT-based peer-to-peer networks. In: Druschel, P., Kaashoek, F., Rowstron, A. (eds.) IPTPS 2002. LNCS, vol. 2429, pp. 242–250. Springer, Heidelberg (2002)
41. Idreos, S., Tryfonopoulos, C., Koubarakis, M.: Distributed evaluation of continuous Equi-join queries over large structured overlay networks. In: ICDE (2006)
42. Palma, W., Akbarinia, R., Pacitti, E., Valduriez, P.: DHTJoin: processing continuous join queries using DHT networks. DPD **26**(2–3), 291–317 (2009)
43. Dédzoé, W.K., Lamarre, P., Akbarinia, R., Valduriez, P.: Efficient early top-k query processing in overloaded P2P systems. In: Hameurlain, A., Liddle, S.W., Schewe, K.-D., Zhou, X. (eds.) DEXA 2011, Part I. LNCS, vol. 6860, pp. 140–155. Springer, Heidelberg (2011)
44. Cai, M., Frank, M., Yan, B., MacGregor, R.: A subscribable peer-to-peer RDF repository for distributed metadata management. J. Web Semant. **2**(2), 109–130 (2004)
45. Liarou, E., Idreos, S., Koubarakis, M.: Continuous RDF query processing over DHTs. In: ISWC (2007)
46. Lohrmann, B., Battré, D., Kao, O.: Towards parallel processing of RDF queries in DHTs. In: Hameurlain, A., Tjoa, A.M. (eds.) Globe 2009. LNCS, vol. 5697, pp. 36–47. Springer, Heidelberg (2009)
47. Battré, D., Heine, F., Höing, A., Hovestadt, M., Kao, O., Liebetruth, C.: Dynamic knowledge in DHT based RDF stores. In: SWWS (2008)
48. Belkin, N., Croft, W.: Information filtering and information retrieval: two sides of the same coin? CACM **35**(12), 29–38 (1992)
49. Li, J., Loo, B., Hellerstein, J., Kaashoek, M., Karger, D., Morris, R.: On the feasibility of peer-to-peer web indexing and search. In: Frans Kaashoek, M., Stoica, I. (eds.) IPTPS 2003, vol. 2735, pp. 207–215. Springer, Heidelberg (2003)
50. Reynolds, P., Vahdat, A.: Efficient peer-to-peer keyword searching. In: Endler, M., Schmidt, D. (eds.) Middleware 2003, vol. 2672, pp. 21–40. Springer, Heidelberg (2003)
51. Hsiao, H.C., King, C.T.: Similarity discovery in structured P2P overlays. In: ICPP (2003)
52. Tryfonopoulos, C., Idreos, S., Koubarakis, M.: LibraRing: an architecture for distributed digital libraries based on DHTs. In: Rauber, A., Christodoulakis, S., Tjoa, A.M. (eds.) ECDL 2005. LNCS, vol. 3652, pp. 25–36. Springer, Heidelberg (2005)
53. Bender, M., Michel, S., Triantafillou, P., Weikum, G., Zimmer, C.: MINERVA: collaborative P2P search (Demo). In: VLDB (2005)
54. Gounaris, A., Fernandes, A., Papadopoulos, A., C. Yfoulis: Parallel query processing on the grid. In: Advances in Parallel Computing (2009)
55. Narendula, R., Papaioannou, T., Aberer, K.: My3: a highly-available P2P-based online social network. In: P2P (2011)
56. Stoica, I., Morris, R., Liben-Nowell, D., Karger, D., Kaashoek, M.F., Dabek, F., Balakrishnan, H.: Chord: a scalable peer-to-peer lookup protocol for internet applications. IEEE/ACM TON **11**(1), 17–32 (2003)

57. Tryfonopoulos, C., Koubarakis, M., Drougas, Y.: Information filtering and query indexing for an information retrieval model. ACM TOIS **27**(2), 1–47 (2009)
58. Yan, T., Garcia-Molina, H.: The SIFT information dissemination system. ACM TODS **24**(4), 529–565 (1999)
59. Huebsch, R.: Content-based multicast: comparison of implementation options. Technical Report UCB//CSD-03-1229, UC Berkeley (2003)
60. Pitoura, T., Ntarmos, N., Triantafillou, P.: Replication, load balancing and efficient range query processing in DHTs. In: Ioannidis, Y. (ed.) EDBT 2006. LNCS, vol. 3896, pp. 131–148. Springer, Heidelberg (2006)
61. Gopalakrishnan, V., Silaghi, B., Bhattacharjee, B., Keleher, P.: Adaptive replication in peer-to-peer systems. In: ICDCS (2004)
62. Shen, H.: Efficient and effective file replication in structured P2P file sharing systems. In: P2P (2009)
63. Deb, S., Linga, P., Rastogi, R., Srinivasan, A.: Accelerating lookups in P2P systems using peer caching. In: ICDE (2008)
64. Bhattacharjee, B., Chawathe, S., Gopalakrishnan, V., Keleher, P., Silaghi, B.: Efficient peer-to-peer searches using result-caching. In: Frans Kaashoek, M., Stoica, I. (eds.) IPTPS 2003, vol. 2735, pp. 225–236. Springer, Heidelberg (2003)
65. Dong, L.: Automatic term extraction and similarity assessment in a domain specific document corpus. Master's thesis, Department of Computer Science, Dalhousie University (2002)
66. Frantzi, K., Ananiadou, S., Mima, H.: Automatic recognition of multi-word terms: the C-value/NC-value method. IJDL **3**(2), 115–130 (2000)
67. Karger, D.R., Ruhl, M.: Simple efficient load balancing algorithms for peer-to-peer systems. In: SPAA (2004)
68. Datta, A., Schmidt, R., Aberer, K.: Query-load balancing in structured overlays. In: CCGRID (2007)
69. Miliaraki, I., Kaoudi, Z., Koubarakis, M.: XML data dissemination using automata on top of structured overlay networks. In: WWW (2008)
70. Kaoudi, Z., Koubarakis, M., Kyzirakos, K., Miliaraki, I., Magiridou, M., Papadakis-Pesaresi, A.: Atlas: storing, updating and querying RDF(S) data on top of DHTs. J. Web Sem. **8**(4), 271–277 (2010)
71. Dean, J., Ghemawat, S.: MapReduce: simplified data processing on large clusters. In: OSDI (2004)

RUBIK: Proactive, Entity-Centric and Personalized Situational Web Application Design

Devis Bianchini[1](✉), Silvana Castano[2], Valeria De Antonellis[1], Alfio Ferrara[2], Elisa Quintarelli[3], and Letizia Tanca[3]

[1] Department of Information Engineering, University of Brescia, via Branze 38,
25123 Brescia, Italy
{bianchin,deantone}@ing.unibs.it
[2] Department of Computer Science, Università degli Studi di Milano,
via Comelico 39, 20135 Milan, Italy
{silvana.castano,alfio.ferrara}@di.unimi.it
[3] Department of Electronics and Information, Politecnico of Milan, via Ponzio 34/5,
20133 Milan, Italy
{quintare,tanca}@elet.polimi.it

Abstract. Over the last years many efforts have been invested in developing *Situational Web Applications (SWAs)*, that is, applications targeted at users' specific requirements. A specific category of SWAs are personalized portals which collect data from documental, social and Semantic Web repositories, often accessed by means of appropriate Web APIs, and present the collected resources tailored on users' needs. Given the growing number and heterogeneity of existing web resources and of the Web APIs to access them, SWA design should be supported by advanced techniques to collect and compose data and Web APIs which are most appropriate for the target users. In light of these considerations, an integrated approach specifically conceived for SWA design should be: (i) *entity-centric*, by clouding data coming from multiple sources related to a given topic of interest; (ii) *personalized*, by filtering data for target users, according to their situations and contexts; (iii) *proactive*, by suggesting Web APIs used to access data of interest in order to ease their composition in the SWAs. In this paper we describe the RUBIK approach, specifically conceived for entity-centric, personalized and proactive composition of SWAs.

Keywords: Web data clouding · Context-aware systems · Personalization · Social web · Linked data · Web APIs · Situational web applications

1 Introduction

Nowadays, the increasing engagement of users in the new generations of webs – like Semantic Web, Web 2.0, Social Web – as both consumers and producers of web resources, has a relevant impact also on the development of Situational Web

A. Hameurlain et al. (Eds.): TLDKS XIII, LNCS 8420, pp. 123–157, 2014.
DOI: 10.1007/978-3-642-54426-2_5, © Springer-Verlag Berlin Heidelberg 2014

Applications [1] (SWAs). SWAs are particular applications targeted at users' specific requirements, like personalized portals which collect data from documental, social and Semantic Web repositories, often accessed by means of suitable Web APIs, and present the collected resources tailored on needs of users acting as resource consumers. Noticeably, the variety of available web resources is the growing, steadily, along with the number and heterogeneity of Web APIs used to access them. On the one hand, unstructured messages, posts, tags, structured data and ontological knowledge are provided and shared among web users. On the other hand, different kinds of domain-specific or general-purpose Web APIs are independently provided by third parties, e.g., to check the availability of hotel rooms, to buy tickets on-line, to display data on a map (e.g., GoogleMaps), to access web resources for Deep-Web-data sharing [2] such as Wikipedia (http://www.mediawiki.org).

In light of these considerations, SWA design should be supported by advanced techniques to collect and compose data and Web APIs which are most appropriate for the target users. For instance, in recent approaches to Web API selection and composition, specifically applied in the context of mashups (which share some features with SWAs), a component is selected on the basis of its past history (e.g., number and type of mashups it has been used in, or co-occurrence with other relevant Web APIs in the same mashups) [3] and on the collective knowledge of other designers who already used, tagged and rated the component on the basis of its use in a particular mashup [4].

These approaches cannot be applied in a straightforward way for modern SWA development. We advocate that a mechanism to identify the topic of interest around which the SWA must be designed and filtering of data according to target users, their situation and context, should support the SWA designer during his/her task. The contribution of this paper is an integrated approach, called RUBIK, overshooting traditional concepts and techniques conceived for SWA design in order to be: (i) *entity-centric*, by clouding data coming from multiple sources related to a given topic of interest; (ii) *personalized*, by filtering data for target users, according to their situations and contexts; (iii) *proactive*, by suggesting Web APIs used to access data of interest in order to ease their composition in the SWAs. In particular, RUBIK applies: (i) *web data clouding* techniques to support the identification of resources related to the topics of interest; such resources are used to better focus the selection of the Web APIs that should be aggregated; (ii) *context-aware and personalized filtering* of the resources identified in the previous phase, exploiting on-the-fly the knowledge about the target users and their context; (iii) *Web API selection and aggregation strategies*, adopted to enable the fast composition of Web APIs used to access resources related to the topics of interest.

1.1   Motivating Example

As a motivating example, consider the case of a web designer, who works for a multimedia and entertainment publishing company and is in charge of designing a personalized portal about the famous movie director Woody Allen. The designer has to collect any kind of data about Allen, using APIs to access various web sources, and to compose them. For instance, the designer may rely on the data already available in various public databases and repositories, like the Internet Movie Database (IMDb), Freebase or Wikipedia. However, once the designer has found these resources, he needs to be supported by a system that can take advantage of all the however shaped information in the web and of its semantics (e.g., ontology knowledge, RSS feeds, news, micro-blogging, SOAP services), to build personalized views over data accessed through available Web APIs. Furthermore, the portal on Woody Allen could be designed for different target users: for a movie critic, who needs a very specialized web site with fresh information for intellectuals and cultivated persons, or for a specialized movie festival organizer, who needs a web site with APIs to find movie theaters and their locations on a map. In both cases, the portal leverages overlapping contents, with common data and Web APIs. The *Situational Web Application Designer* (hereafter, the *SWADler*) needs a support system which minimizes the effort spent to design the portal, by working on a common collection of relevant Web APIs and data, and properly filtering them to present only the most suitable/pertinent portion(s) for a specific target user.

Modern approaches on Web API selection and composition implement advanced recommendation strategies based on collective knowledge on Web API use in the past [3], combined with semantic tagging and search techniques [4], properly weighting the expertise of other designers in Web application development [5]. Those approaches provide relevant results from which our research may start. Nevertheless, these solutions do not enable the SWADler to:

– perform Web API and resource selection starting from a cloud-link scheme; tag clouds are an effective visual mean, widely applied in the Social Web context, to display any possible kind of data and data relationships following a folksonomy-like style; a tag cloud-link scheme can be used to gather data related to a topic of interest, coming from multiple webs and browsing them according to an exploratory perspective, that best fits the condition of a SWADler, who may have not exactly in mind what are the contents more suitable to be included in the application being developed; existing approaches do not provide more than a keyword-based search or a search based on semantic tagging, using the "page view" access style of search engines and focusing on Web API descriptions, not considering also web data on which the Web APIs provide access; moreover, data clouds may provide keyword expansion which increases the recall of Web API and web resource searching process;
– filter the available data, given a topic of interest, on the basis of the knowledge of the target users and their current situation; this can be performed tailoring the cloud-link scheme on users' contents and needs; existing context-aware Web application design methodologies [6] are not devoted to the selection of third-party Web APIs and do not start from available data clouds; nevertheless, context-aware filtering may significantly increase the precision of the Web API and web resource searching process.

In the following, we will discuss how the distinctive features of the RUBIK approach, that is, the joint application of data clouding, context-aware web resource filtering and Web API selection, may be exploited to meet the SWADler's requirements introduced above.

1.2 Paper Organization

This paper is organized as follows. The next section describes the reference methodology and the support tool components characterizing the RUBIK approach. Sections 3, 4, 5 and 6 are the core sections of the paper, devoted to describe in details the RUBIK phases. Section 7 is devoted to evaluation issues while Sect. 8 reviews the state-of-the-art to highlight the cutting-edge elements of RUBIK. Finally, Sect. 9 closes the paper.

2   The RUBIK Approach

The RUBIK approach for SWA design fosters a reference methodology together with envisaged versions of tool support workflows to help the SWA designer during his/her task where API should be realized.

2.1   The Proposed RUBIK Methodology

The RUBIK methodology is shown in Fig. 1.

Web data clouding is the first phase, to identify, classify and organize relevant web resources into web data clouds. Starting from a target of interest which synthesizes in form of keyword(s) the goal of the SWA to be designed, the web data clouding phase produces a large collection of web resources about the target, extracted from documental, social and Semantic Web repositories. Extracted resources are organized into a web data cloud graph structure, where each node represents a *cluster* of similar web resources and an edge represents a relation of *proximity* between clusters. Clusters are associated with Web APIs used to access the underlying web resources. Techniques for the construction of the web data cloud will be described in Sect. 3.

The second phase of the SWA design concerns the context-driven filtering of the web data cloud. The goal is to identify possible personalizations within the web data cloud (for subsequent selection of web resources and APIs) for the construction of different SWAs depending on the needs of their target users. To this aim, the RUBIK system builds an internal model of the target users' contexts in order to prune the web data cloud to produce *contextual data clouds*. A contextual data cloud is actually a different *facet* on the web data cloud – therefore the name RUBIK for our system, suggesting a cube with several faces on the relevant contents. The internal context model captures all the possible contexts the target users of the SWA might be acting in, by means of a set of dimensions such as: the main topics of interest, the position, the current user role such as movie critic or movie festival organizer. The detailed description of

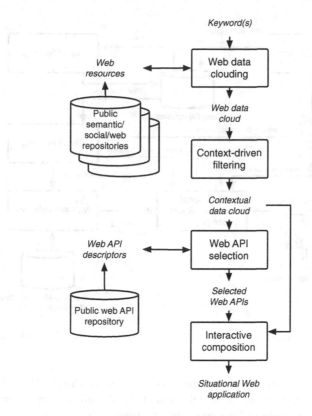

Fig. 1. The RUBIK methodology for SWA design.

the context model on which this phase relies and its exploitation for personalized SWA design will be described in Sect. 4.

The third and fourth phases rely on advanced techniques to select Web APIs apt to access data for the target of interest and to proactively support the SWADler during an incremental SWA composition by highlighting: (i) APIs that are similar each other and can be included as alternatives in the SWA under construction; or (ii) APIs that can be easily included together in the SWA being developed. Each time the SWADler inserts a new Web API, the system suggests APIs similar to those already included in the SWA and additional APIs that can be easily coupled with other APIs already included in the application. This interactive Web API composition is performed by relying on similarity between Web APIs extracted from the `ProgrammableWeb` repository and their co-occurrences in past Web mashups, which can also be inferred from the same repository. We chose the `ProgrammableWeb` repository since, to the best of our knowledge, it is the most popular and updated Web API repository available on-line. Section 5 will detail the way the Web APIs are properly selected starting from the contextual data cloud. Web API interactive composition will be described in Sect. 6.

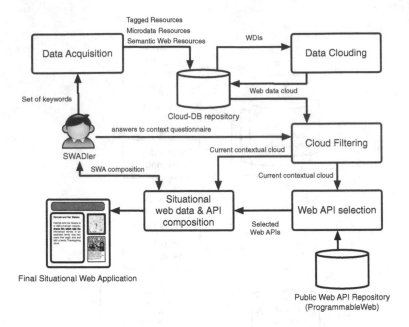

Fig. 2. The functional modules of the general-purpose version of RUBIK.

The four phases ensure the three innovative aspects of the RUBIK approach: web data clouding enables entity-centric collection of web resources and related APIs around a given topic of interest; context-driven filtering ensures personalization of SWA contents; finally, the last two phases enable proactive exploration and selection of available data and Web APIs from huge and heterogeneous API repositories.

2.2 RUBIK Versions

The proposed methodology is implemented by the functional modules depicted in Figs. 2 and 3, corresponding to two different versions of the RUBIK system. Each version captures a typical situation of SWA design, as follows:

- a *general-purpose version* (see Fig. 2), which is domain-independent, in that it supports entity-centric, personalized and proactive SWA design in different application domains of interest; working with this version, keyword(s) specified by the SWADler for the topic of interest are used to lookup all the repositories of the various webs in order to extract relevant resources to build the web data cloud through the *Data Acquisition* module; moreover, the SWADler defines the context of the target users of the SWA by answering a predefined questionnaire to assign values to the contextual dimensions; such values refer to generic contextual perspectives as they are not related to a specific application domain;

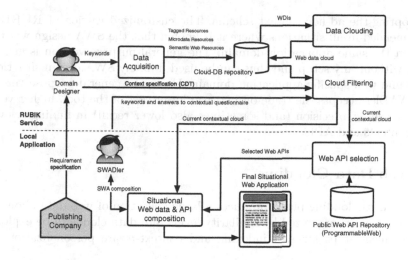

Fig. 3. The functional modules of the customized version of RUBIK.

– a *customized version* (see Fig. 3), which is domain-focused, in that it supports
  SWADlers during the SWA design activities within a certain domain of interest
  (e.g., movies from the 1970s to the 1990s); the customized version comes with
  a pre-defined web data cloud built at the setup stage with the help of a
  specialized domain designer; the domain designer is a domain expert who
  interacts with the SWADler to gather all information about the application
  domain the SWADler is interested in, in order to setup a comprehensive web
  data cloud and a focused context schema; working with the customized version
  means that the SWADLer keyword(s), specifying the target of interest (i.e.,
  Woody Allen), are directly applied to the pre-defined domain web data cloud
  (i.e., the movies from the 1970s to the 1990s), thus making the definition of
  the web data cloud for the target much more efficient than in the case of
  the general-purpose version; in the pre-defined web data cloud, all potentially
  useful resources have been included by the domain designer, such as the copies
  of the movies themselves, the biographies of their actors and directors, the
  books that inspired them and similar; moreover, the contextual dimensions
  instantiated by the SWADler have been also defined by the domain designer
  having in mind the specific application domain, thus they are more focused
  than in the general-purpose version.

In both versions, Web API selection and interactive composition over the contextual data cloud proceed in a similar way. The *Web API selection* module supports
the extraction of available Web APIs from the `ProgrammableWeb` repository, after
which the *Situational web data & API composition* module proactively supports
the SWADler in the aggregation of the selected APIs to obtain the final personalized SWA. Using the customized version, the SWADler is able to build more
precise and customized SWAs, by acting on the pre-defined web data cloud and

by adopting the ad-hoc context schema. The customized version of RUBIK is recommended in all situations where it is known that the SWA design will take place in the same application domain. The general-purpose version is generic, comes without any setup and thus can be used in many SWA design situations, not bound to a specific application domain. Using the general-purpose version, the SWADler accepts the trade-off between lower cost of the tool, higher versatility and lower precision (and sometimes also lower recall) in finding relevant Web contents and APIs.

3 Web Data Clouding

The web data clouding phase produces a large collection of web data about the target of interest, organized by similarity levels into data clouds to be exploited in subsequent steps for API indexing and context-aware personalization and filtering.

3.1 Data Acquisition

RUBIK works on different kinds of web resources: (i) *tagged resources*, that are web resources coming from bookmarking and social annotation systems, including annotated web pages, images and videos; (ii) *microdata resources*, that are web resources coming from microblogging systems and news feeds; (iii) *semantic web resources*, that are web resources coming from RDF(S) knowledge repositories, Linked Data sources and/or OWL ontologies.

The goal of data acquisition is to represent and store each acquired web resource in a homogeneous way, based on the notion of web data item (WDI in the following). A web data item wdi_i is a metadata representation of a web resource i, such as for example a single annotated web page, an RSS news item, or an RDF/OWL individual. WDIs represent also properties and types in the case of semantic web resources (e.g., RDF/OWL properties and classes, Freebase types). A WDI is denoted by a unique identifier, a human-readable name and a resource type, whose values, either I, P, or C, denote the fact that the WDI represents an individual, a property, or a class/type, respectively. WDIs are associated with a collection of terms extracted from the original web resource (called *terminological equipment*). In particular, the terminological equipment is extracted from tags in case of tagged resources, from the textual content of microdata resources and from the property values of semantic web resources. Moreover, in case of semantic web resources, WDIs are associated with a set of types, taken from the classification of the resource in the original datasource, and with a set of predicates, which represent the properties of the resources and their values.

WDIs are stored into the cloud-DB repository, namely a relational database designed according to the schema shown in Fig. 4. The cloud-DB repository is conceptually divided into three main sections: (i) data collection, storing WDI representations of all the web resources collected for a target of interest; (ii) data

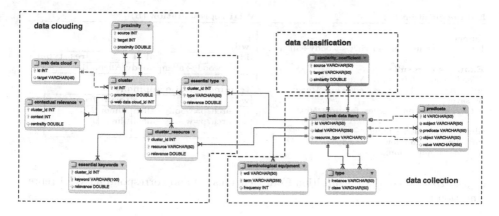

Fig. 4. Schema of the cloud-DB repository.

classification, storing the output of the classification process in terms of similarity results; (iii) data clouding, storing the data cloud structure. In particular, for what concerns data collection, the cloud-DB provides a main table (wdi), which contains WDIs featured by their readable name (a label) and the resource type. Each WDI is then associated with its terminological equipment (terminological equipment table). Each term in the terminological equipment is associated with the number of times the term appears in the original resource (i.e., frequency). The type equipment of a WDI is represented through the type table. When we acquire a resource that has a type (e.g., an ontological instance, an RDF description), we create in the wdi table a WDI of type instance for the resource and other WDIs of type class for representing the original types/classes of the resource. Then, the type table is used in order to store the association between instances and types. Moreover, a WDI is also associated with one or more predicates (predicate), that are featured by a property, a concrete value and an abstract value. The property is another WDI used to present properties of the original datasource. The concrete value is used to store original property values consisting in strings, dates, numbers. The abstract value is used to store the property value when it consists of a reference to another WDI.

Data are acquired by means of specific wrappers designed on top of the structure of the web data sources selected for the acquisition process, by using the APIs provided by the web source at hand, when available. In the general-purpose version of RUBIK, pre-defined wrappers are available for a set of web sources, which actually include Twitter (http://twitter.com), Delicious (http://delicious.com/), a generalized wrapper for the RSS 2.0 standard, Freebase (http://www.freebase.com/) and DBpedia (http://dbpedia.org). In the customized version of RUBIK, when new specific web sources are required for SWA design, the corresponding wrappers can be added to the data acquisition module according to a modularized architecture, which requires only to create the wrapper and connect it to the cloud-DB for storing the WDIs acquired from the web source at hand.

Data acquisition (a)	WDI representation (b)

Query
http://delicious.com/search?p=woody+allen

WDI representation (b)

wdi

id	label	res. type
...?p=woody+allen	Woody Allen...	I

Example of results
Woody Allen Returns to New York
in 'Whatever Works' – New York Magazine
39 saves
http://nymag.com/movies/features/56930/

film – woodyallen – humor – movies
larrydavid – interview – comedy
article – funny – humour

terminological equipment

wdi	term	frequency
...search?p=woody+allen	humor	2
...search?p=woody+allen	film	1
...	...	...

Fig. 5. Example of data acquisition from Delicious (a) and corresponding WDI representation (b).

In order to provide examples of data acquisition supported in RUBIK, we consider the Delicious repository, the IMDb movie web site and Freebase.

An example of data acquired from Delicious by submitting the query "Woody Allen"[1] is shown in Fig. 5, together with its corresponding WDI representation.

As shown in the example, a WDI is created for each entry of Delicious, reporting the URL of the tag list and the entry title as label. Then, each tag is associated with the newly created WDI in the terminological_equipment table, together with the number of occurrences of each tag.

An example of data acquisition from the IMDb RSS channel about Allen is shown in Fig. 6. WDI representation of an RSS news item (and, in general, of microdata resources) is characterized by a set of properties featuring the resource itself. The set of properties is extracted from the flat structure of microdata resources, such as, in the case of RSS news items, the title and the publication date. These properties are represented in RUBIK as predicates associated with the WDI. The terminological equipment is built by extracting a list of featuring terms from the textual content of the resource.

Data acquisition from Freebase is performed by first retrieving the Freebase ID /en/woody_allen corresponding to Woody Allen and subsequently running a set of queries expressed through the MQL query language used in Freebase, such as the one shown in Fig. 7. The MQL query shown in Fig. 7 searches for all the movies directed by Woody Allen. The WDI representation extracted from the query answer is defined according to the RDF graph that can be derived from the model (e.g., RDF(S)/OWL) associated with the resource. The terminological equipment is defined as the set of all the terms appearing in the nodes and edges of the RDF graph such as concept names, property names, URI labels, comments and literals. Predicates are built by adding a new WDI predicate for each edge of the RDF graph. WDI types are defined by taking into account the classification of the resource in the RDF graph, by adding a new type for each RDFS class associated with the resource at hand.

---

[1] http://delicious.com/search?p=woody+allen

Data acquisition (a)	WDI representation (b)

Query
http://www.imdb.com/news/ni26786922/

Example of results
Film Review: 'To Rome with Love'
22 April 2012 3:36 PM, PDT
Having outstayed his welcome in London
before a brief and pleasant sojourn in Paris,
Woody Allen continues his European tour,
this time landing in Rome with a heavy
(handed) bump courtesy of latest release
To Rome with Love (2012)...

wdi

id	label	res. type
...ni26786922/	Film Review: 'To Rome...	I
title	title	P
pubDate	date	P

terminological equipment

wdi	term	frequency
...ni26786922/	european	1
...ni26786922/	rome	2
...	...	...

predicate

id	subject	predicate	object	value
149	...ni26786922/	title	NULL	Film Rev...
150	...ni26786922/	pubDate	NULL	22/04/2012
...	...	...	...	...

Fig. 6. Example of data acquisition from IMDb news (a) and corresponding WDI representation (b).

3.2 Data Clouding

A web data cloud is defined as a graph $\mathcal{G} = (\mathcal{N}, \mathcal{E})$, where a node $n_i \in \mathcal{N}$ represents a *cluster* of web resources and an edge $e_j(n_i, n_k) \in \mathcal{E}$ represents a relation of *proximity* between clusters n_i and n_k, respectively.

The construction of a web data cloud is based on a classification process, which has the goal of clustering similar web resources by exploiting information in their WDI representations. A cluster in a web data cloud contains a set of similar web resources among those collected for the target. A web data cloud cluster is associated with an *essential*, that is a synthetic and representative description of the cluster contents. The essential of a cluster cl_i is composed of a set K_i of *keywords* and a set T_i of *types* extracted from terminological and type equipments of the WDIs belonging to cl_i, respectively. In particular, K_i contains the most relevant terms in the terminological equipments of WDIs belonging to cl_i, while T_i contains the most relevant types in the type equipments of WDIs belonging to cl_i. Term and type relevance values for a cluster cl_i are calculated using conventional TF/IDF Information Retrieval measures [7].

A proximity relation between two clusters denotes the fact that resources therein contained are in some way related. Proximity relations are labeled with a degree of proximity, that is a measure of the strength of the relation holding between involved clusters. In particular, given two clusters cl_i and cl_j, their proximity relation represents a relation of content-based similarity between cl_i and cl_j and it is associated with a proximity degree X_{ij} which is proportional to the number of similar WDIs between cl_i and cl_j over the number of all the WDIs in cl_i.

Finally, a cluster cl_i in a web data cloud is characterized by a degree of *prominence* which measures the relative relevance of cl_i in the web data cloud considering the number and strength of proximity relations holding between cl_i

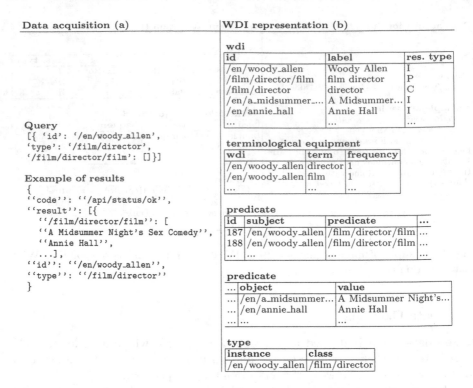

Fig. 7. Example of data acquisition from Freebase (a) and corresponding WDI representation (b).

and the other clusters of the web data cloud. In RUBIK, prominence is computed on the basis of the random walks procedure that has been proposed in [8]. This measure is calculated by counting how often a cluster cl_i is traversed by a random walk between two other clusters, using proximity relations between clusters as paths in the web data cloud.

An example of web data cloud about the target "Woody Allen" is shown in Fig. 8. In the example, clusters are represented as circles whose diameter is proportional to the cluster prominence. Proximity relations between clusters are represented as lines whose thickness is proportional to the degree of proximity. Clusters are associated with their corresponding essentials, represented as squares. Finally, each cluster is also associated with a sample of web resources it contains, shown in a dashed squares.

To construct the web data cloud, we employ matching techniques to evaluate the level of similarity between the WDIs stored in the cloud-DB repository. The choice of the matching techniques to use has to comply with the different complexity which characterizes the various web resources, and consequently, their corresponding WDI representations. We note that the terminological equipment is the only equipment always defined, despite of the kind of web resource. Moreover, for a tagged or a microdata resource, the terminological equipment

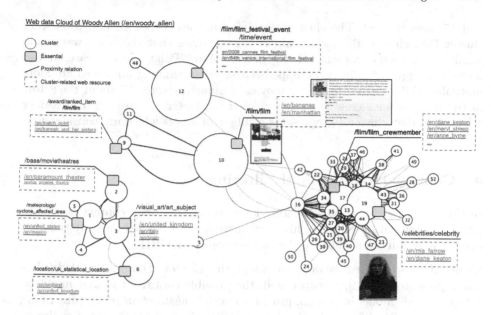

Fig. 8. Graphical representation of the ten most prominent clusters in the web data cloud for Woody Allen.

captures most of the informative content of the whole resource. For this reason, term matching techniques play a crucial role for similarity evaluation. Moreover, term matching techniques are exploited by more articulated matching techniques to evaluate type and structural measures of semantic similarity between web resources. WDI similarity evaluation is performed by exploiting the matching library of HMatch 2.0 matching system [9,10], where a wide set of string matching functions are available to accommodate different matching requirements and cases.

On the basis of similarity coefficients resulting from matching, a *hierarchical* clustering technique of *agglomerative* type is employed to perform WDIs classification [11,12]. "Agglomerative" refers to the property of the technique to proceed by a series of successive merging of similar WDIs into groups. "Hierarchical" refers to the property of the technique to classify WDIs into groups at different levels of similarity to form a similarity tree, where leaves represent WDIs and internal nodes correspond to virtual elements (called "centroids"). Given a similarity tree, clusters of similar WDIs are selected according to a threshold-based mechanism. In particular, given a threshold t, each cluster cl_i selected for clouding corresponds to the largest subtree of T whose root node (i.e., the similarity coefficient of cl_i) is greater than or equal to t. With respect to the web data cloud about Woody Allen shown in Fig. 8 the agglomerative clustering process produces several clusters representing movies, actors and other subjects related to the activity of Allen as a director. As an example, we can focus on clusters representing movies and actors working in Allen movies (i.e., clusters 10 and 17, respectively). The cluster containing movies is then related to another cluster (i.e., cluster 9) which is about those movies that received awards and nominations, such as for example the movie "Match Point". In this case, movies have been aggregated in two groups because those winning an award are more mutually similar than other Allen movies. Similarly, cluster 17 about the actors in the Allen movies crew is strictly related to cluster 19 through a proximity relation, in that cluster 19 represents a subset of actors which are considered as celebrities.

4   Context-Driven Cloud Filtering

In this section we show how the knowledge about the user's context is used to prune the web data cloud. For this purpose we have developed an early-tailoring approach, equally applicable to the two versions of RUBIK.

Context Model. In this work we adopt the *Context Dimension Model*, formally presented in [13], to capture all the possible contexts the user might be acting in. This model is *general*, provides *multiple abstraction levels* (describing the contexts with different levels of granularity), is *readable* (to serve as design documentation) and *expressive* (allowing querying, reasoning or constraint specification on the contexts) [14,15]. The Context Dimension Model provides the constructs to design the *context schema* (or *Context Dimension Tree – CDT*) by means of a hierarchical structure. Figure 9 shows (a) the CDT built by the domain designer for the customized version of RUBIK, in the case of our running example, and (b) the generic CDT, built-in into the general-purpose version of RUBIK. The CDT consists of (i) *context dimensions* (black nodes), modeling the context variables, that is the different perspectives through which the user perceives the application domain (e.g., *situation*, *interest_topic*) and (ii) the allowed *dimension values* (white nodes), i.e., the possible values based on which the contexts are to be instantiated.

The CDT of Fig. 9(a) has as dimensions the *situation*, the current *interest-topic*, the *event* and the *genre* of movies the user is interested in, the *time* and *location*. Each dimension node has as children the possible values the domain designer considers relevant for the specific target domains. For example, the *interest-topic* values are: *movie*, *actor*, *director*. Each of such values has a parameter which can be instantiated at run-time in case the SWADler wants to choose a specific movie (e.g., mID), actor (e.g., aID) or director of interest (e.g., dID).

The CDT of Fig. 9(b) has not been defined by the domain designer. Each of the dimensions *domain* and *interest_topic* assumes as value a set of keywords: these are used, together with spatial and temporal information (the two other dimensions), to filter the web data cloud.

Any set of dimension values from the CDT can be thus used to filter the web data cloud information. According to a formal point of view, a context is defined as a conjunction of *context elements*, i.e., statements of the form

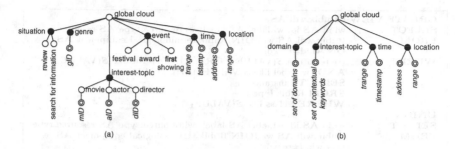

Fig. 9. The Context Dimension Trees for the running example, in the customized version (a) and general-purpose version (b) of RUBIK.

dimension=value, and represents a possible point of view we used to regard the relevant cloud portion.

Let us consider the customized version of RUBIK for our running scenario related to movies from the 1970s to the 1990s. The SWADler wants to create a SWA to look for information about awards related to dramatic movies directed by Woody Allen. His target context, derived from the CDT of Fig. 9(a), is: $C_1 \equiv genre=(gID=$ *"dramatic"*$) \land event=$ *"award"*$\land$ *interest_topic*$= director(dID=$ *"Woody Allen"*$)$.

In the general-purpose version of RUBIK, instead, the SWADler has to develop a SWA for a movie critic writing a review about movies by Woody Allen. The target context derived from the CDT of Fig. 9(b) is $C_2 \equiv set\_of\_domains=$ *"movies"* $\land$ *set_of_contextual_keywords*$=$ *"Woody Allen"*. Note that, in this case, the SWADler needs to specify the domain "movies", which is implicit in the customized version of RUBIK because of the preparatory work of the domain designer.

The adoption of a hierarchical structure allows us to employ different abstraction levels to specify and represent contexts. Indeed, even if this is not highlighted in our running example, a dimension value can be further refined by using (sub-)dimensions. It is also possible to define appropriate constraints on the schema, which prevent meaningless combinations of context elements for the current application scenario. Moreover, the CDT provides support to context evolution. We do not delve into details here; for a formal and complete description refer to [16].

The SWADler, who wants to build a personalized SWA, fills out a predefined questionnaire to determine the context of the target users of the application. By focusing the attention on relevant information for the possible active contexts, the semi-automatic tailoring of web data cloud can be performed. In the customized version of RUBIK, at design time the SWADler must associate to each context a view over the web data cloud. In the general-purpose version, this association is also hard-wired in RUBIK. In both cases, the association remains *virtual*, and will be applied to the whole cloud, so that the SWADler will be

```
CREATE VIEW CRubik.wdi AS
SELECT       wi.id AS id, wi.label AS label, wi.resource_type AS resource_type
FROM         Rubik.wdi AS wi JOIN Rubik.terminological_equipment AS te
             ON wi.id=te.wdi
WHERE        (te.term like S(VALUE₂) OR ... OR te.term like S(VALUEₙ))
             AND (wi.label like S(VALUE₁) OR wi.id in
             SELECT t.instance
             FROM Rubik.type t
             WHERE t.class like S(VALUE₁))
UNION
SELECT       wi.id AS id, wi.label AS label, wi.resource_type AS resource_type
FROM         Rubik.wdi AS wi JOIN Rubik.terminological_equipment AS te
             ON wi.id=te.wdi
WHERE        (te.term like S(VALUE₁) OR ... OR te.term like S(VALUEₙ))
             AND (wi.label like S(VALUE₂) OR wi.id in
             SELECT t.instance
             FROM Rubik.type t
             WHERE t.class like S(VALUE₂))
UNION
...
```

Fig. 10. The generic SQL view to tailor the `wdi` relation.

provided access only to Web APIs useful to build the SWA coherent with the target context.

Definition of the contextual views. Once the CDT for the specific application domain has been defined, to associate the various possible contexts with the context-relevant portions of the web data cloud, the SWADler specifies *contextual views* over the cloud-DB database (see the schema in Fig. 4). The contextual view for each context C is defined as a set of SQL views, one for each relation R_i of the database, that can be used to: (i) pick the entire relation R_i, (ii) filter some tuples by specifying a selection condition and/or by using only the join operator with other relations, (iii) combine different expressions on R_i by means of the intersection or union operators.

In previous work [13] the context-aware view on a database for a context C is manually written as a set of *relational algebra expressions* over the database schema. In RUBIK, where the cloud-DB database is known, contextual views are defined by using the SQL language in an almost automatic way, since the conditions for the WHERE clause of SQL queries can be extracted from the keywords and the context specification is obtained from the questionnaire.

Specifically, given a generic context

$$C \equiv (dim_1 = VALUE_1) \wedge \cdots \wedge (dim_n = VALUE_n) \tag{1}$$

a contextual view filters the data collection part of the cloud-DB database, and in particular the `wdi` relation (see Fig. 10).

As shown in Fig. 10, this SQL view is composed as the union of n subqueries, where n is the number of context elements composing the context C. The query contains the operator $S(\text{keyword}) = \{k_1, \ldots, k_n\}$, which is the synset (that is, the set of synonyms) for the concept "keyword" returned by WordNet lexical system [17]. The values $VALUE_i$ are obtained from Eq. (1), further automation

```
CREATE VIEW CRubik.wdi AS
SELECT        wi.id AS id, wi.label AS label, wi.resource_type AS resource_type
FROM          Rubik.wdi AS wi JOIN Rubik.terminological_equipment AS te
              ON wi.id=te.wdi
WHERE         (te.term LIKE S("award") OR te.term LIKE S(" Woody Allen"))
              AND (wi.label LIKE S("dramatic") OR wi.id in
              SELECT t.instance
              FROM Rubik.type t
              WHERE t.class LIKE S("dramatic"))
UNION
SELECT        wi.id AS id, wi.label AS label, wi.resource_type AS resource_type
FROM          Rubik.wdi AS wi JOIN Rubik.terminological_equipment AS te
              ON wi.id=te.wdi
WHERE         (te.term LIKE S("dramatic") OR te.term LIKE S( "Woody Allen"))
              AND (wi.label LIKE S("award") OR wi.id in
              SELECT t.instance
              FROM Rubik.type t
              WHERE t.class LIKE S("award"))
UNION
SELECT        wi.id AS id, wi.label AS label, wi.resource_type AS resource_type
FROM          Rubik.wdi AS wi JOIN Rubik.terminological_equipment AS te
              ON wi.id=te.wdi
WHERE         (te.term LIKE S("dramatic") OR te.term LIKE S("award"))
              AND (wi.label LIKE S("Woody Allen") OR wi.id in
              SELECT t.instance
              FROM Rubik.type t
              WHERE t.class LIKE S(" Woody Allen"))
```

Fig. 11. The SQL view to tailor the `wdi` relation for context C_1.

can be achieved by using the dimension values extended with their synsets. Each subquery in Fig. 10 applies the join operator between the relations `wdi` and `terminological_equipment`. The subquery extracts from the `wdi` relation: (a) the tuples having the value of the `label` attribute that matches one of the words in the synset of $VALUE_i$ and contains a term that matches one of the other dimension values; or (b) the tuples related to a tuple in the `type` relation having the value of the `class` attribute that matches the same condition.

For example, to build a SWA for the context C_1, the system will impose a filter on the table `wdi` based on the synsets of the *genre* "dramatic" and the *interest_topic* value "award", then on the synset of the *director* "Woody Allen". The resulting SQL query is shown in Fig. 11.

Cloud filtering: generation of the contextual data clouds. A contextual data cloud for a target context C is obtained by considering, from the original web data cloud, only those clusters that are related to the WDIs in the contextual view `CRubik.wdi`, that is:

```
CREATE VIEW C1Rubik.cluster AS
SELECT        wi.id AS id, wi.label AS label, wi.resource_type AS resource_type
FROM          Rubik.cluster AS c , Rubik.essential_type AS et,
              C1Rubik.wdi AS wi
WHERE         cr.cluster_id=c.id AND c.id=et.cluster
              AND c.id=et.cluster_id AND et.type=wi.id
```

At this point, the RUBIK system uses the contextual data cloud to provide the SWADler only with the APIs related to the contextual data cloud essentials. In the customized version of RUBIK, during the process of virtual tailoring, the SWADler might judge that, for certain contexts, the contextual data clouds produced automatically by RUBIK are not satisfactory. For example, they might be "too large". RUBIK provides the SWADler with the possibility to modify the automatically produced views. In this case, in order to further reduce the number of clusters and consider only those that are really important, he/she restricts the view on the relation CLUSTER on the basis of the centrality value stored in the `contextual_relevance` relation. For context C_1 the contextual view becomes:

```
CREATE VIEW  C1Rubik.cluster as
SELECT       wi.id AS id, wi.label AS label, wi.resource_type AS resource_type
FROM         Rubik.cluster AS c , Rubik.essential_type AS et,
             C1Rubik.wdi AS wi, Rubik.contextual_relevance AS cr
WHERE        cr.context=C_1 AND cr.centrality> μ AND cr.cluster_id=c.id
             AND c.id=et.cluster AND c.id=et.cluster_id AND et.type=wi.id
```

In this way, the contextual data cloud for the context C_1 contains only the clusters that are both related to WDIs selected for C_1 and in the context C_1 have a centrality value greater than a threshold μ. Note that these design choices make the contextualization of the customized version much more accurate than in the general-purpose case.

The contextual data cloud generated from the views for the context C_1 is shown in Fig. 12. Note that some clusters of the original web data cloud have been excluded: only 25 clusters out of 53 have been selected in the tailoring process. In this case, no filter has been applied to the centrality value. For example, in John's contextual data cloud the cluster number 2 describing the locations where movies are shown is not present, since this information is not relevant in the current context. The tuples in the `wdi` relation are 332 (they were 12546 in the original table) because only WDIs related to the "award" concept have been selected. Globally, the contextual data cloud is reduced to 25 % of the original web data cloud; the tailoring process has improved both the focus of the clusters and related WDIs and their quality; indeed, also a great number of numeric values not related to terms mentioning the concept "award" (or one of the concepts in its synset) are not included in the contextual data cloud.

It is interesting to note that the use of context can be coupled with that of the SWADlers' personal preferences, thus allowing for a finer personalisation. In general, contextual preferences determine a ranking of the data based on the actual interests and needs of a user when she is in a particular context [18]. Adopting this technique here would induce a ranking on web cloud data as well as the set of recommended APIs. However, since asking the SWADlers to manually specify all their preferences seems unrealistic, it is possible to automatically infer their interests through data mining, adopting the technique described in [19]. The inference step can be performed by collecting a log of the SWADlers' choices (in terms of querying and other activity) when they act in each of the possible

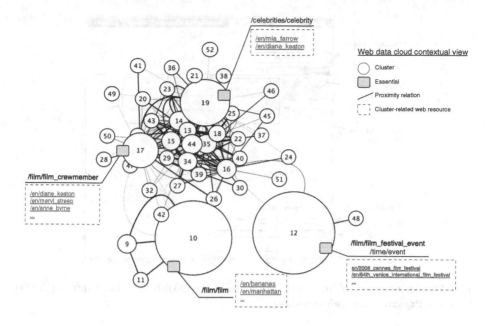

Fig. 12. Graphical representation of the clusters in the contextual data cloud of C_1.

contexts, and learn from these choices in order to (dynamically) configure a ranked list of web data (and APIs) to be offered next time the same context arises.

5   Web API Selection

The RUBIK system provides advanced functionalities to collect Web API descriptions, link them to the contextual data cloud elements, and evaluate their similarity to enable more effective, cloud-driven Web API selection and aggregation. The Web API selection module relies on Web API records within the `ProgrammableWeb` repository, which provides the proper methods to retrieve them[2].

The goal of Web API collection is to represent and store Web API Descriptors (WADs). A WAD $\mathcal{W}_i$ is a metadata representation of a Web API in terms of a name, one or more categories $(C_{\mathcal{W}_i})$, a Uniform Resource Identifier (URI) and a human-readable description. The WAD is also associated with a collection of tags $\{t_{\mathcal{W}_i}\}$ *(Web API terminological equipment)*. Moreover, the WAD is related to a set $\mathcal{M}_{\mathcal{W}_i}$ of one or more existing mashups, which include the Web API represented by the WAD. Each mashup is in turn described by a name, a URI, a

---

[2] `api.programmableweb.com/`.

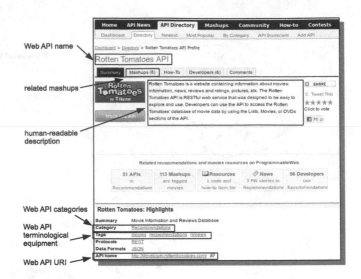

Fig. 13. An example of information used to extract Web API Descriptors (WADs) from the `ProgrammableWeb` repository.

human-readable description and the set of WADs of Web APIs which the mashup is composed of. All the information about the WADs and the related mashups are extracted from the `ProgrammableWeb` repository. In order to provide an example of Web API Descriptor, we may take into account the `Rotten Tomatoes` Web API to retrieve information about movies (see Fig. 13); its WAD is the following:

$\mathcal{W}_1 =$ [NAME: Rotten Tomatoes API;

URI: http://developer.rottentomatoes.com/;

DESCRIPTION: Rotten Tomatoes is a website containing information about movies...;

CATEGORIES: {Recommendations};

TERMINOLOGICAL EQUIPMENT: {movies, recommendations, reviews};

MASHUPS: $\mathcal{M}_{\mathcal{W}_1}$]

According to the `ProgrammableWeb` repository, the `Rotten Tomatoes API` has been used in eight mashups. Among them, the `MovieGram` application is designed for finding informations, trailers and ratings on movies, directors and actors and contains also the `Trailer Addict` and `YouTube API`; the `InstantPlex` application is designed to watch trailers, movies and TV shows and contains eight Web APIs:

$m_1{}^1 \in \mathcal{M}_{\mathcal{W}_1} =$ [NAME: MovieGram;

URI: http://moviegr.am/;

DESCRIPTION: A quick and easy way to find movies, watch trailers and share with friends!;

WADS: {Rotten Tomatoes, Trailer Addict, YouTube}]

$m_1{}^2 \in \mathcal{M}_{\mathcal{W}_1} =$ [NAME: InstantPlex;

    URI: http://instantplex.com/;

    DESCRIPTION: Discover, queue and watch trailers, movies and TV shows...;

    WADS: {Facebook, Freebase, GetGlue, Netflix, Rotten Tomatoes, The Movie DB, Twitter, YouTube}]

The activity of linking Web APIs to the contextual data cloud is performed through the identification of matching between the set K_i of keywords and the set T_i of types of the essential of each cluster cl_i in the contextual data cloud and the terminological equipment of each Web API as extracted from the ProgrammableWeb repository. The matching between the terminological equipment of Web API descriptors and the features of essentials in the contextual data cloud is evaluated by applying state-of-the-art term matching techniques also used for web data clouding acquisition and classification during the cloud construction [9].

5.1   Web API Similarity Evaluation

Web API similarity is evaluated on the WADs. Specifically, similarity between WADs can be exploited during web application development to substitute a Web API already included in the application with a similar one as described in the next section. Web APIs can be substituted because: (i) they become unavailable; (ii) application requirements have been changed; (iii) the application context is changed. Similarity between WADs is evaluated by considering different aspects, namely the *category similarity* $Sim_c()$, the *tag similarity* $Sim_t()$ and the mashup similarity $Sim_m()$. We will explain them with the help of the following example. Let us consider the MoviePilot Web API, to retrieve information about movies, whose WAD is the following one:

$\mathcal{W}_2 =$ [NAME: MoviePilot;

    URI: http://code.google.com/p/moviepilot-api/;

    DESCRIPTION: Moviepilot is a site which shows the latest news about movies...;

    CATEGORIES: {Recommendations};

    TERMINOLOGICAL EQUIPMENT: {search, related, recommendations, movies};

    MASHUPS: $\mathcal{M}_{\mathcal{W}_2}$]

According to the ProgrammableWeb repository, the MoviePilot Web API has been used in the following mashup only:

$m_2{}^1 \in \mathcal{M}_{\mathcal{W}_2} =$ [NAME: VIDVIDOO;

    URI: http://vidvidoo.com/;

    DESCRIPTION: VIDVIDOO New Releases and Classics Movie Reviews...;

    WADS: {Amazon eCommerce, MoviePilot, Netflix, Rotten Tomatoes, The Movie DB, YouTube}]

Category similarity. The similarity between two categories c_i and c_j is inferred from the ProgrammableWeb repository. Since no hierarchies are defined among the available categories, advanced semantic-driven techniques (such as category subsumption checking) can not be used. Nevertheless, we consider the two categories as more similar as the number of Web APIs that are categorized in both the categories increases with respect to the overall number of Web APIs classified in c_i and c_j. The average similarity between two WADs $\mathcal{W}_1$ and $\mathcal{W}_2$ based on their categories, denoted with $Sim_c(\mathcal{W}_1, \mathcal{W}_2) \in [0,1]$, is computed, through the application of the Dice formula [7], as the average similarity between pairs of categories, one from $C_{\mathcal{W}_1}$ and one from $C_{\mathcal{W}_2}$. Pairs of categories to be considered in the Sim_c computation are selected according to a maximization function relying on the assignment in bipartite graphs, which has been introduced in [20]. This function ensures that each category from $C_{\mathcal{W}_1}$ participates in at most one pair with one of the categories from $C_{\mathcal{W}_2}$ and viceversa and the pairs are selected in order to maximize the overall Sim_c.

For instance, let us consider the Rotten Tomatoes and MoviePilot APIs, whose WADs have been described above. They are classified in only one category, Recommendations, which is the same for both of them and contributes to the category similarity between $\mathcal{W}_1$ and $\mathcal{W}_2$ with 1.0. According to the Dice formula, the category similarity is therefore computed as:

$$Sim_c(\mathcal{W}_1, \mathcal{W}_2) = \frac{2 \cdot (1.0)}{2} = 1.0 \qquad (2)$$

where the numerator is doubled since we are comparing two WADs and the denominator in the formula represents the total number of categories in $C_{\mathcal{W}_1}$ and $C_{\mathcal{W}_2}$.

Tag similarity. The similarity between two WADs $\mathcal{W}_1$ and $\mathcal{W}_2$ based on their tags, denoted with $Sim_t(\mathcal{W}_1, \mathcal{W}_2) \in [0,1]$, is computed by evaluating the term affinity between pairs of tags, one from the terminological equipment of $\mathcal{W}_1$ and one from the terminological equipment of $\mathcal{W}_2$, and by combining them through the Dice formula. Also in this case, pairs are selected according to the same maximization function used for category similarity evaluation. The term affinity between two tags t_1 and t_2 belongs to the range $[0,1]$ and is computed as extensively described in [20], based on WordNet. In WordNet, synsets are related by eighteen different kinds of relationships. In particular, *hyponymy/hypernymy relations* are used to represent the specialization/generalization relationship between two terms: for instance, movie is a more specific term with respect to show; this means that there is a semantic affinity between movie and show. According to this viewpoint, the affinity between two tags t_1 and t_2 is maximum (that is, equal to 1.0) if the tags belong to the same synset or coincide; otherwise, if they belong to different synsets, a path of hyponymy/hypernymy relations which connects the two synsets is searched: the highest the number of relationships in this path, the lowest the term affinity.

For instance, if we consider the terminological equipment of Rotten Tomatoes and MoviePilot APIs, whose WADs have been described above, there are two common tags (movies and recommendations) which contribute to the tag similarity between $\mathcal{W}_1$ and $\mathcal{W}_2$ with $1.0 + 1.0$ (out of $3 + 4 = 7$ tags in the two terminological equipments). The tag similarity is therefore computed as:

$$Sim_t(\mathcal{W}_1, \mathcal{W}_2) = \frac{2 \cdot (1.0 + 1.0)}{7} = 0.571 \tag{3}$$

where the total term affinity is doubled since we are comparing two WADs.

Mashup similarity. Similarity between two mashups, denoted with $MashSim()$, is computed as the number of common Web APIs used in both mashups with respect to the overall number of Web APIs used in the two mashups. For instance, if we consider the mashups $m_2{}^1$ and $m_1{}^2$ in the running example, we have:

$$MashSim(m_2{}^1, m_1{}^2) = \frac{2 \cdot 4}{8 + 6} = 0.571 \tag{4}$$

since the two mashups share `Netflix`, `Rotten Tomatoes`, `The Movie DB` and `YouTube` APIs out of $8 + 6 = 14$ Web APIs. Similarly, $MashSim(m_1{}^1, m_1{}^2) = 0.444$. The average similarity between two WADs $\mathcal{W}_1$ and $\mathcal{W}_2$ based on the mashups where they have been included, denoted with $Sim_m(\mathcal{W}_1, \mathcal{W}_2) \in [0, 1]$, is computed by applying the Dice formula to the values of $MashSim()$ between pairs of mashups, one from $\mathcal{M}_{\mathcal{W}_1}$ and one from $\mathcal{M}_{\mathcal{W}_2}$, using the same maximization function adopted for the other kinds of similarity evaluation. For instance, in the running example, since $MashSim(m_1{}^1, m_1{}^2) < MashSim(m_2{}^1, m_1{}^2)$, we select the second pair[3]. $\mathcal{W}_1$ has been included in eight mashups, while $\mathcal{W}_2$ has been included in only one mashup, therefore

$$Sim_m(\mathcal{W}_1, \mathcal{W}_2) = \frac{2 \cdot 0.571}{8 + 1} = 0.127 \tag{5}$$

6   Interactive Composition

Coming now to the SWA design, the SWADler starts his interaction by introducing the keyword "Woody Allen" and some information about the target context, by means of the ad-hoc questionnaire. Then, RUBIK presents the SWADLer not only with the portion of web data cloud filtered on the basis of keyword "Woody Allen" and current context information, but also with the available Web APIs related to this fragment of cloud, gathered by using the introduced keyword and context information, and the elements contained in the cloud about Allen. The APIs can be of different nature. There could be search APIs to obtain information about Allen's books (e.g., Amazon.com) or movies (e.g., IMDb), or blogs collecting discussions, comments and opinions about books or movies. Now the SWADler can drag&drop the proposed Web APIs in the design canvas area and build its SWA. Figure 14 shows the interface presented to a SWADler.

The RUBIK system proactively suggests step-by-step the Web APIs to select and to aggregate in the application being developed. We distinguish between *completion suggestions* and *substitution suggestions*. In the former, the SWADler looks for other Web APIs that could be added to the ones already put in the

---

[3] In this example, we did not show the *mashup closeness* of the other mashups in $\mathcal{M}_{\mathcal{W}_1}$ with respect to $m_2{}^1$ because they are all below the $MashSim(m_2{}^1, m_1{}^2)$ value.

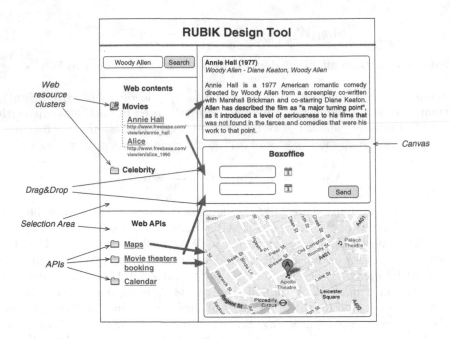

Fig. 14. The RUBIK interface.

canvas. In the latter, the SWADler looks for Web APIs that are similar to one of the Web APIs already put in the canvas, that we denote with $\mathcal{W}_s$, properly selected by the SWADler on the canvas, in order to replace it.

For the *Web API completion suggestions*, the RUBIK system considers all the mashups that include at least one of the Web APIs already put in the canvas. RUBIK evaluates, for each available WAD $\mathcal{W}_i$ included in those mashups and not yet selected for the composition, the percentage of mashups that include $\mathcal{W}_i$. This percentage is used to rank the Web APIs proposed to the SWADler. For instance, let us consider the design of a SWA where the `Rotten Tomatoes` API only has been selected and put in the canvas. The `Netflix` Web API has been used in five out of eight mashups, where the `Rotten Tomatoes` API has been used[4] (see Fig. 15). Therefore, `Netflix` is proposed as a first suggestion to the designer.

For the *Web API substitution suggestions*, the Web API similarity evaluation described in the previous section is taken into account. In particular, we distinguish two cases: (1) the SWADler has just started the development of a new situational application, that is, the canvas only contains the Web API $\mathcal{W}_s$ to be substituted; (2) the canvas contains more than one Web API (and, among them, it contains $\mathcal{W}_s$). If we are in the first case, the problem is to find Web APIs $\mathcal{W}_i$ in the repository such that the linear combination of their category, tag and

---

[4] For a complete list of these mashups, see http://www.programmableweb.com/api/
rotten-tomatoes/mashups.

Web APIs	Mashups where it has been used together with the Rotten Tomatoes API
YouTube	4 of 8
Rovi Cloud Services, Freebase, The Movie DB, Trailer Addict	2 of 8
Netflix	5 of 8
Facebook, GetGlue, Twitter, Last.fm, Kynetx, Amazon eCommerce, MoviePilot, Fb Social Plugins	1 of 8

Fig. 15. List of APIs used in one of the eight mashups that include also the Rotten Tomatoes API.

Compared APIs	Sim_c	Sim_t	Sim_m	Average similarity
$\mathcal{W}_1$ and $\mathcal{W}_2$	1.0	0.571	0.127	0.566
$\mathcal{W}_1$ and $\mathcal{W}_3$	1.0	1.0	0.0	0.667

Fig. 16. Similarity values to suggest the substitution of the Rotten Tomatoes API ($\mathcal{W}_1$) with the MoviePilot API ($\mathcal{W}_2$) or the Filmaster API ($\mathcal{W}_3$) when the canvas only contains Rotten Tomatoes.

mashup similarity with $\mathcal{W}_s$, equally weighted, is different from zero. In this case, such a linear combination is also used to rank the suggested Web APIs $\mathcal{W}_i$. For instance, let us imagine to have the $\mathcal{W}_1$ Web API only on the canvas. Moreover, let's consider the $\mathcal{W}_2$ Web API in the example above and the following Web API Descriptor:

$\mathcal{W}_3 = [$NAME: Filmaster;

    URI: http://filmaster.org/display/DEV/API;

    DESCRIPTION: Movie review and recommendation service;

    CATEGORIES: {Recommendations};

    TERMINOLOGICAL EQUIPMENT: {movies, recommendations, reviews};

    MASHUPS: $\mathcal{M}_{\mathcal{W}_3}]$

According to the ProgrammableWeb repository, the Filmaster API has not been used in any other mashup. In this example, the similarity values are shown in Fig. 16. Therefore, the $\mathcal{W}_3$ API is suggested first to the SWADler.

If we are in the second case, the suggestion of an alternative Web API $\mathcal{W}_i$ must take into account the category, tag and mashup similarity of $\mathcal{W}_i$ with $\mathcal{W}_s$, but also the number of times the other Web APIs in the canvas have been used in the same mashups where $\mathcal{W}_i$ has been included. For instance, let us suppose now that the canvas contains the Rotten Tomatoes and Netflix APIs and the MoviePilot and Filmaster APIs are considered for substituting the Rotten Tomatoes API. According to the ProgrammableWeb repository, when MoviePilot is used, also Netflix API is included in the same mashup, while the Netflix API and the Filmaster API have not been ever used in the same mashup. Therefore, the similarity values and percentages shown in Fig. 17 follow and the MoviePilot API is ranked first.

Compared APIs	Sim_c	Sim_t	Sim_m	% of times the second Web API has been used together with Netflix	Linear combination of values
$\mathcal{W}_1$ and $\mathcal{W}_2$	1.0	0.571	0.127	1.0 (100%)	0.674
$\mathcal{W}_1$ and $\mathcal{W}_3$	1.0	1.0	0.0	0.0 (0%)	0.5

Fig. 17. Similarity values to suggest the substitution of the Rotten Tomatoes API ($\mathcal{W}_1$) with the MoviePilot API ($\mathcal{W}_2$) or the Filmaster API ($\mathcal{W}_3$) when the canvas contains Rotten Tomatoes and the Netflix API.

7   Evaluation Issues

The RUBIK approach merges two fundamental aspects, namely (i) *web data filtering* by means of web data clouding and context-aware pruning, and (ii) *proactive Web API selection and aggregation*. In this respect, there are neither benchmarks to compare the RUBIK system with similar efforts nor universally accepted evaluation parameters on which the comparison can be based. As a consequence, we have chosen to rely on generally agreed-upon information retrieval and software engineering metrics to evaluate:

1. the scalability of the web data filtering step to find only relevant data for a given context (aspect (i));
2. the effectiveness and scalability of the Web API selection step to find relevant Web APIs (aspect (ii));
3. the quality of the RUBIK system perceived by different kinds of users who are supported during the construction of a personalized SWA (aspect (ii)).

All the experiments have been performed on an Intel laptop, with a 2.53 GHz Core 2 Due CPU, 2GB RAM and Linux operating system. Each experiment was run ten times and the experimental results show the average. In all cases, the highest deviation from the average was not more than 3%.

7.1   Web Data Filtering

The web data clouding activity naturally raises some scalability issues, also taking into account that the way the data clouds are built differs in the two versions of RUBIK. In particular, the keyword-driven approach to data acquisition can potentially lead to the extraction of large collections of web data items. Working with the general-purpose version, such large collections must be managed online by the SWADLer, thus affecting system performance and scalability. In the case of the customized version, large pre-defined domain data clouds are built off-line, thus data clouding scalability is not a big concern for system performance. For scalability evaluation in data cloud construction, we observe that scalability is mainly affected by the employed clustering techniques. In fact, matching, that is the other time-consuming activity for data cloud construction, has been optimized in our matching system HMatch [10] by relying on a state-of-the-art technique for comparison reduction in case of large collections of items to match [21]. Concerning clustering and its scalability, we executed tests with the hierarchical clustering algorithm, and we run a set of experimental tests on different datasets in the movie domain extracted from Delicious, IMDb and Freebase, containing a growing number of web data items, ranging from 30 up to 5000. Time complexity of hierarchical and agglomerative clustering is $O(n^2)$, where n is the number of items to be clustered, and problems of scalability of this approach are well known. However, our experimental results show that the approach scales quite well when the number of items involved is lower than 5000, by requiring an average execution time lower than 1s. This is a reasonable result for many kinds of user requests, considering that data extraction queries can be configured to acquire a focused number of items in order to achieve a faster computation of the web data cloud. However, also when large collections of data items are extracted, scalability can be achieved through strategies where matching and clustering are performed off-line (e.g., in a batch manner), like in the customized version, or through caching mechanisms to exploit previous data clouding results.

The contribution of the contextualization (tailoring) phase affects scalability and performance as much as any relational database query. This phase works on-line in both versions of RUBIK, when the SWADler composes the application. Contextual views automatically generated, as shown in Sect. 4, are Unions of Conjunctive Queries (UCQs), which are known to have LOGSPACE complexity [22]. The cloud filtering described in the same section presents the same complexity. However, just as in traditional databases, the most frequently used views – corresponding to the most incurred contexts – can be pre-computed off-line, thus yielding shorter or null on-line waiting time.

7.2   Web API Selection and Aggregation

To evaluate the effectiveness and scalability of the RUBIK system for Web APIs suggestion, we ran a set of experiments focusing on the application domain of the running example. In the experiments, we compared different systems and different kinds of search: (i) keyword-based Web API search performed on the ProgrammableWeb repository; (ii) keyword-based Web API search performed on the ProgrammableWeb repository after the expansion of keyword set through the application of the web data clouding; (iii) Web API search performed with the support of the RUBIK system. Results are presented in Fig. 18. To evaluate the precision (i.e., the fraction of suggested Web APIs that are relevant) and recall of the search results (i.e., the fraction of relevant Web APIs that are suggested by the system), a domain expert manually selected a set of 124 relevant Web APIs out of the ProgrammableWeb repository, classified within the Entertainment, Events, Mapping, Media Management, Recommendations and Video categories. Manual Web API selection has been performed by analysing WADs and, starting from them, by analysing all the Web APIs included in the related mashups. This enables to identify the most common functionalities/features included in existing applications for the considered domain. For example, starting from the `Rotten Tomatoes` Web API, the associated mashups (such as `MovieGram` and `InstantPlex`) have been analysed to identify other Web APIs commonly included in this kind of applications. Figure 18(a) shows the precision and recall values when no Web APIs have been selected and put in the canvas yet, but a single keyword has been specified (e.g., `movie`, `director` or `actor` in the running example). As expected, the recall increases, due to the keyword expansion given by the web data clouding. Moreover, also precision increases, due to the application of the context-aware filtering. Precision and recall values for experiments performed in the substitution scenario, when only one Web API has been put in the canvas and must be substituted, are very similar to the ones shown in Fig. 18(a); in particular, in this test we performed search on the `ProgrammableWeb` repository using as keywords the tags in the terminological equipment of the Web API to be substituted. Precision and recall values change significantly on the `ProgrammableWeb` repository when more than one API have been put in the canvas and one of them has to be substituted (see Fig. 18(b)), while the behavior of RUBIK remains almost unchanged. In Fig. 18(b) we compared our system also against ApiHut [23], which relies on a classification of Web APIs based also on other features extracted from the `ProgrammableWeb` repository, such as the protocol used by the Web APIs or the data format. The ApiHut solution presents good precision, but recall values decrease with respect to RUBIK due to the absence of keyword set expansion performed in our system through the web data clouding phase. Figure 18(c) shows the precision and recall values for experiments performed in the completion scenario. In this case, we did not compare RUBIK against keyword-based search performed on the `ProgrammableWeb` repository and against the ApiHut approach, since they have not been designed for proactive suggestions in this kind of scenario. On the other hand, we compared RUBIK against MatchUp [3], whose suggestions rely on existing Web mashups stored in the `ProgrammableWeb` repository. The precision of MatchUp decreases with respect to our system, since we considered a wider range of similarity evaluations which makes the RUBIK suggestions more effective, while the MatchUp system relies on existing mashups only. The recall values are comparable.

We also ran an additional set of tests, where we used the keyword "Woody Allen" (that is, a keyword representing an instance in the domain of interest) to start the Web data clouding. The use of this keyword directly on the `ProgrammableWeb` repository does not produce any result. The web data clouding enables the retrieval of the set K_i of keywords and the set T_i of types of the essentials in the (contextual) data cloud that increases the precision and recall of both the search with the expanded set of keywords on the `ProgrammableWeb` repository and of the search performed on the RUBIK system (see Fig. 18(d)).

In the RUBIK approach, scalability issues can also be due to Web API similarity evaluation during Web API selection phase. We extracted WADs from the `ProgrammableWeb` repository and we plotted similarity evaluation response time with respect to the number n of WADs (Fig. 19). The time required for the Web API linking anf the computation of similarities as shown in Eqs. (2)–(5) is

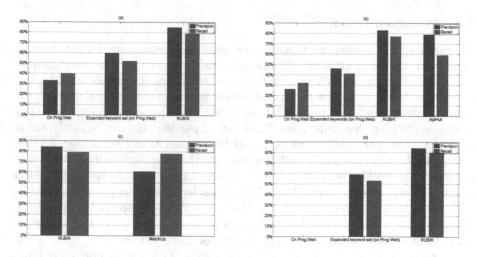

Fig. 18. Precision and recall values during Web API selection with RUBIK system compared against other approaches.

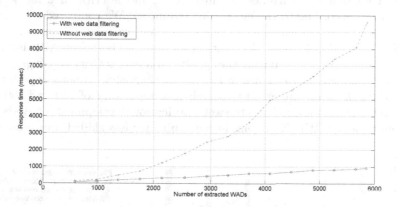

Fig. 19. Scalability of the Web API similarity evaluation for Web API selection and aggregation purposes.

negligible due to the application of optimized term matching techniques defined in [9] between small sets of tags from the Web API terminological equipments and due to the optimized computation of the Dice formula as shown in [20]. Therefore, the complexity of Web API similarity evaluation is dominated by the number n of Web APIs to be compared against the request. Specifically, the complexity increases almost linearly with respect to n (see Fig. 19). The web data clouding phase and the subsequent context-aware filtering phase enable to filter out not relevant Web APIs, considering only Web APIs that are related to the target of interest. This makes the Web API selection phase more scalable.

7.3 Perceived Quality

We ran a set of experiments to assess the quality of personalized SWA development with the support of the RUBIK system. In particular, we monitored six users for building new SWA with Web APIs of the running example considered in this paper. Users 1, 4 and 6 have middle-level web application development skill, while users 2, 3 and 5 present high-level web application development skill. It is worth mentioning that using a small number of participants can be motivated as in [24,25], where the authors propose a mathematical model about the effectiveness of usability experiments and demonstrate that running multiple tests with a small number of users is more effective than running a single test with a large number of users. We performed ten Web API selection and aggregation tests for each user, by increasing both the number of available Web APIs (from 10 to 500) and the complexity of the application to build, in terms of the number of Web APIs to select (from 2 to 5), for a total of 60 experiments.

In each experiment, we considered a set of Web APIs for the domain of interest, whose WADs have been extracted from the `ProgrammableWeb` repository (see Sect. 7.2), as well as not relevant Web APIs and we asked users to build a new SWA for different purposes (for instance, a movie critic or a movie festival organizer, according to the motivating example). We compared the number of correct and wrong answers of six users with and without the support of the RUBIK system (C_S, W_S, C_N, W_N) with respect to the best Web APIs indicated by the domain expert, computing the quality of answers $Q(x) = \frac{C_x}{C_x + W_x}$.

The average results are shown in Fig. 20, distinguishing among experimental results when the users are asked to search in a repository of 250 and 500 Web APIs. We note that the quality of the web application development increases with the support of the RUBIK system, in particular for less skilled users and when

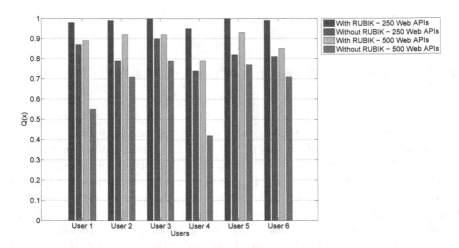

Fig. 20. Average quality of personalized SWA development with and without RUBIK support.

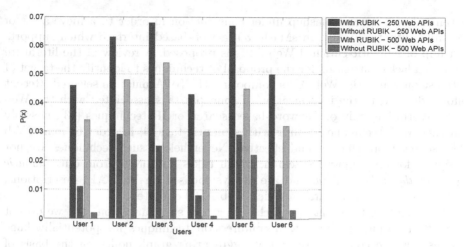

Fig. 21. Average productivity of personalized SWA development with and without RUBIK support.

the number of available APIs is high, thus requiring much effort for selecting the right ones. Even when the quality without RUBIK strongly decreases – with 500 available APIs with respect to the case of 250 available APIs (see, for instance, users 1 and 4) – the support of RUBIK enables to reach high quality levels.

If we consider the productivity to develop a personalized SWA, defined as $P(x) = \frac{C_x}{t_x}$, where t_x is the time (in seconds) taken for answering, we obtain a more detailed measure of how much the RUBIK support impacts on the tasks of SWADlers. Figure 21 displays the average productivity to build personalized SWAs, starting from 250 and from 500 available APIs.

8 Related Work

Recent efforts to support fast development of web applications underline the importance of mitigating the burden of Web API selection and aggregation. Specifically, in [23] a faceted classification of Web APIs and a ranking algorithm to improve their retrieval are proposed, based on IR techniques that are applied to Web API information only, as published on the `ProgrammableWeb` repository. The approach described in [3] relies on existing web mashups stored in the repository. In [26] semantic annotations have been proposed to enrich Web API modeling and proper metrics based on such annotations have been defined to improve recommendations on Web API retrieval and aggregation. Exploratory search is applied directly on Web API registry browsing in [27–30]. In particular, in [27] a tag-based navigation technique for composing data mashups is proposed. The MashMaker system [28] suggests Web APIs that might assist in handling data currently managed by the web designer (for instance, the tool might suggest adding "map" location or "distance" APIs if the designer currently views a list of addresses). The MashupAdvisor system [29] computes an extension of a specific mashup under construction in order to achieve a set of possible outputs. Finally, in sMash [30] a web-based interface which supports mashup of semantic-enriched Web APIs is proposed. A review of the literature showed a lack of attention for the proposal of techniques to identify the target of interest on which the Web APIs composing the SWA must be selected. Recent efforts devoted to the identification of concepts/tags associated with the Web API descriptions rely on keyword-based search or IR techniques [23], possibly enriched with techniques dealing with semantic tags [4] in order to make Web API search more efficient and effective. Nevertheless, such techniques are not designed for an explorative search, which takes advantage from our *thematic view on the webs of data*, and are always focused on Web API descriptions, instead of targeting at data on which Web APIs operate.

With respect to recent approaches to graph summarization [31], we do not aim at providing efficient graph compression techniques for potentially huge graphs. On the contrary, we aim at aggregating graph nodes on the basis of their mutual relations and their similarity. The ultimate goal of our approach is to aggregate graph nodes with respect to a theme, in order to simplify the exploration of the linked data cloud, up to a final, intuitive, and easy-understandable data cloud made of few *essentials* equipped with *prominence* and *proximity* information. For what concerns approaches for search and retrieval of information coming from the different webs, their goal is moving from traditional information lookup to exploratory search. Here exploratory search is defined as the activity of finding and understanding knowledge about a topic of interest by exploiting aggregation and learning of information in a social context [32]. With respect to these approaches, we stress the role of data similarity, proximity and prominence as basic techniques. Therefore, our approach constitutes a step forward with respect to, for example, Sig.ma (Semantic Information MAshup) [33], which retrieves and integrates linked data, starting from a single URI, by querying the Web of Data and applying machine learning to the data found. Moreover, in RUBIK we go beyond these solutions by providing a linking towards available Web APIs to be selected and aggregated on top of the `ProgrammableWeb` repository.

On the top of the thematic clusters, the current designer's context works as a viewpoint mechanism which takes into account implicit or explicit background knowledge. This prunes the available data and APIs coming from the web, thus enabling personalization in SWA development. Sophisticated and general context models have been proposed [6,14], to support Web service retrieval [34] or *knowledge chunks*, determining the set of situation-relevant information [35,36] and services [37,38]. In this case, context-aware modeling aims at completing existing models for Web API selection and aggregation, which are mainly focused on the representation of the Web APIs and of their composition [39,40], taking into account semantic annotation of Web API elements for their effective retrieval [41]. Moreover, in RUBIK the data structures employed for cloud representation and the strong semantic grounding of the approach allow to push the automatization of the tailoring of the contextual data cloud further than in previous work [36], which needed a heavier intervention of the designer. In fact, for the generation of the possible contexts, in RUBIK the domain designer intervenes in the case of the customized version, while the general-purpose version relies on a set of generic context dimensions whose values are derived from the SWADler's answers to the questionnaire. For what concerns the generation of contextual data clouds, the stable nature of the internal relational schema storing the cloud data suggests a fixed, generic structure for the contextual chunks, whose design does not require the domain designer intervention.

9 Conclusions

In this work we introduced the RUBIK approach for the personalized composition of webs to satisfy information and resource-delivery application requirements in different contexts. An experimental evaluation related to the movie domain has also been presented, showing the effectiveness of the approach. The practical impact of the functionalities provided by the RUBIK system is relevant in different domains. Besides the web design purpose discussed above, another situation that would immediately benefit from the same on-the-fly integration approach is data integration for life-science researchers, mostly based on web resource discovery and integration. Existing data integration systems (e.g., Swiss Prot) are very complex and do not have the flexibility of accepting requests for heterogeneous resources. Hence the community has still the need to access constantly varying sources (dynamicity), potentially on a large scale, therefore becoming an ideal target of the RUBIK system.

References

1. Kraiem, N., Selmi, S., Ghezala, H.: A situational approach for web applications design. Int. J. Comput. Sci. Issues **7**, 37–51 (2010)
2. Wright, A.: Exploring a 'Deep Web' That Google Can't Grasp, http://www.nytimes.com/2009/02/23/technology/internet/23search.html?pagewanted=all. Accessed: May 2012. The New York Time (February 2009)
3. Greenshpan, O., Milo, T., Polyzotis, N.: Autocompletion for mashups. In: Proceedings of the 35th International Conference on Very Large DataBases (VLDB'09), Lyon, France, pp. 538–549 (2009)
4. Bianchini, D., De Antonellis, V., Melchiori, M.: Semantic collaborative tagging for web APIs sharing and reuse. In: Brambilla, M., Tokuda, T., Tolksdorf, R. (eds.) ICWE 2012. LNCS, vol. 7387, pp. 76–90. Springer, Heidelberg (2012)
5. Bianchini, D., De Antonellis, V., Melchiori, M.: A multi-perspective framework for web API search in enterprise mashup design. In: Salinesi, C., Norrie, M.C., Pastor, Ó. (eds.) CAiSE 2013. LNCS, vol. 7908, pp. 353–368. Springer, Heidelberg (2013)
6. Baldauf, M., Dustdar, S., Rosenberg, F.: A survey on context-aware systems. Int. J. Ad Hoc Ubiquit. Comput. **2**(4), 263–277 (2007)
7. van Rijsbergen, C.J.: Information Retrieval. Butterworth, London (1979)
8. Newman, M.J.: A Measure of betweenness centrality based on random walks. Soc. Netw. **27**(1), 39–54 (2005)

9. Castano, S., Ferrara, A., Montanelli, S., Varese, G.: Ontology and instance matching. In: Paliouras, G., Spyropoulos, C.D., Tsatsaronis, G. (eds.) Multimedia Information Extraction. LNCS, vol. 6050, pp. 167–195. Springer, Heidelberg (2011)

10. Castano, S., Ferrara, A., Montanelli, S.: Matching ontologies in open networked systems: techniques and applications. In: Spaccapietra, S., Atzeni, P., Chu, W.W., Catarci, T., Sycara, K. (eds.) Journal on Data Semantics V. LNCS, vol. 3870, pp. 25–63. Springer, Heidelberg (2006)

11. Castano, S., De Antonellis, V., De Capitani di Vemercati, S.: Global viewing of heterogeneous data sources. IEEE Trans. on Knowl. Data Eng. **13**(2), 277–297 (2001)

12. Castano, S., Ferrara, A., Montanelli, S.: Structured data clouding across multiple webs. Inf. Syst. **37**(4), 352–371 (2012)

13. Bolchini, C., Quintarelli, E., Tanca, L.: CARVE: context-aware automatic view definition over relational databases. Inf. Syst. **38**, 45–67 (2013)

14. Bolchini, C., Curino, C.A., Quintarelli, E., Schreiber, F.A., Tanca, L.: A data-oriented survey of context models. SIGMOD Rec. **36**(4), 19–26 (2007)

15. Mileo, A., Merico, D., Bisiani, R.: Support for context-aware monitoring in home healthcare. In: Intelligent Environments (Workshops), pp. 177–184 (2009)

16. Quintarelli, E., Rabosio, E., Tanca, L.: Context schema evolution in context-aware data management. In: Jeusfeld, M., Delcambre, L., Ling, T.-W. (eds.) ER 2011. LNCS, vol. 6998, pp. 290–303. Springer, Heidelberg (2011)

17. Fellbaum, C.: Wordnet: An Electronic Lexical Database. MIT Press, Cambridge (1998)

18. Miele, A., Quintarelli, E., Tanca, L.: A methodology for preference-based personalization of contextual data. In: Proceedings of EDBT 2009, 12th International Conference on Extending Database Technology, pp. 287–298 (2009)

19. Miele, A., Quintarelli, E., Rabosio, E., Tanca, L.: A data-mining approach to preference-based data ranking founded on contextual information. Inf. Syst. **38**(4), 524–544 (2013)

20. Bianchini, D., Antonellis, V.D., Melchiori, M.: Flexible semantic-based service matchmaking and discovery. World Wide Web J. **11**(2), 227–251 (2008)

21. Euzenat, J., Shvaiko, P.: Ontology Matching. Springer, Heidelberg (2007)

22. Vardi, M.Y.: The complexity of relational query languages (extended abstract). In: STOC, pp. 137–146 (1982)

23. Gomadam, K., Ranabahu, A., Nagarajan, M., Sheth, A., Verma, K.: A faceted classification based approach to search and rank web APIs. In: Proceedings of International Conference on Web Services (ICWS 2008), Beijing, China, pp. 177–184 (2008)

24. Nielsen, J.: Why You Only Need to Test with 5 Users (2000), http://www.useit.com/alertbox/20000319.html Accessed: May 2012

25. Nielsen, J., Landauer, T.: A mathematical model of the finding of usability problems. In: Proceedings of the INTERACT '93 and CHI '93 Conference on Human Factors in Computing Systems, pp. 206–213 (1993)

26. Bianchini, D., Antonellis, V.D., Melchiori, M.: Semantic-driven mashup design. In: Proceedings of 12th International Conference on Information Integration and Web-based Applications and Services (iiWAS'10), pp. 245–252 (2010)

27. Riabov, A., Boillet, E., Feblowitz, M., Liu, Z., Ranganathan, A.: Wishful search: interactive composition of data mashups. In: Proceedings of the 19th International World Wide Web Conference (WWW'08), Beijin, China, pp. 775–784 (2008)

28. Ennals, R., Garofalakis, M.: MashMaker: Mashups for the Masses. In: Proceedings of the 27th ACM SIGMOD International Conference on Management of Data, pp. 1116–1118 (2007)
29. Elmeleegy, H., Ivan, A., Akkiraju, R., Goodwin, R.: MashupAdvisor: a recommendation tool for mashup development. In: Proceedings of 6th International Conference on Web Services (ICWS'08), Beijin, China, pp. 337–344 (2008)
30. Lu, B., Wu, Z., Ni, Y., Xie, G., Zhou, C., Chen, H.: sMash: semantic-based mashup navigation for data API network. In: Proceedings of the 18th International World Wide Web Conference, pp. 1133–1134 (2009)
31. Tian, Y., Hankins, R., Patel, J.: Efficient aggregation for graph summarization. In: Proceedings of the 2008 ACM SIGMOD International Conference on Management of Data, pp. 567–580. ACM (2008)
32. Marchionini, G.: Exploratory search: from finding to understanding. Commun. ACM **49**(4), 41–46 (2006)
33. Tummarello, G., et al.: Sig. ma: live views on the web of data. Web Semant.: Sci., Serv. Agents World Wide Web **8**(4), 355–364 (2010)
34. Raverdy, P.-G., Riva, O., de La Chapelle, A., Chibout, R., Issarny, V.: Efficient context-aware service discovery in multi-protocol pervasive environments. In: Mobile Data Management, p. 3. IEEE Computer Society (2006)
35. Roussos, Y., Stavrakas, Y., Pavlaki, V.: Towards a context-aware relational model. In: Proceedings of 1st International Context Representation and Reasoning, Work, pp. 7.1–7.12 (2005)
36. Bolchini, C., Curino, C.A., Quintarelli, E., Schreiber, F.A., Tanca, L.: Context information for knowledge reshaping. Intl J. Web Eng. Technol. **5**(1), 88–103 (2009)
37. Raverdy, P., Riva, O., de La Chapelle, A., Chibout, R., Issarny, V.: Efficient context-aware service discovery in multi-protocol pervasive environments. In: Proceedings of 7th International Conference on Mobile Data Management, pp. 3–11 (2006)
38. Gu, T., Pung, H.K., Zhang, D.Q.: A service-oriented middleware for building context-aware services. J. Netw. Comput. Appl. **28**(1), 1–18 (2005)
39. Cappiello, C., Matera, M., Picozzi, M., Sprega, G., Barbagallo, D., Francalanci, C.: DashMash: a mashup environment for end user development. In: Auer, S., Díaz, O., Papadopoulos, G.A. (eds.) ICWE 2011. LNCS, vol. 6757, pp. 152–166. Springer, Heidelberg (2011)
40. Abiteboul, S., Greenshpan, O., Milo, T.: Modeling the mashup space. In: Proceedings of the Workshop on Web Information and Data Management, pp. 87–94 (2008)
41. Bianchini, D., De Antonellis, V., Melchiori, M.: Semantics-enabled web API organization and recommendation. In: De Troyer, O., Bauzer Medeiros, C., Billen, R., Hallot, P., Simitsis, A., Van Mingroot, H. (eds.) ER 2011 Workshops. LNCS, vol. 6999, pp. 34–43. Springer, Heidelberg (2011)

Mining Multiple Related Data Sources
Using Object-Oriented Model

C.I. Ezeife$^{(\boxtimes)}$ and Dan Zhang

School of Computer Science, University of Windsor, Windsor, ON N9B 3P4, Canada
{cezeife,woddlab}@uwindsor.ca
http://cezeife.myweb.cs.uwindsor.ca/

Abstract. An object-oriented database is represented by a set of classes connected by their class inheritance hierarchy through superclass and subclass relationships. An object-oriented database is suitable for capturing more comprehensive and detailed complexity of real world data such as capturing multiple related tables representing data schemas of a retail store web site, or capturing multiple databases such as several retail store web sites. Modeling web and other data as a number of object database schemas would enable derived, historical, and comparative mining of multiple databases and tables.

This paper proposes an object-oriented class model and database schema, and a series of class methods including that for object-oriented join (OOJoin) for mining multiple data sources through object oriented model. The OOJoin procedure joins superclass and subclass tables by matching their type and super type relationships. Mining Hierarchical Frequent Patterns (MineHFPs) from multiple integrated databases is done by applying an extended TidFP technique which specifies the object class hierarchy by traversing the multiple database inheritance hierarchy. This paper also extends map-gen join method used in TidFP algorithm to oomap-gen join for generating k-itemset object candidate patterns. The oomap-gen join reduces the number of candidate itemsets generated through indexing of the (k-1)-itemset candidate pattern with start and end position codes for the inheritance hierarchy level. Experimental results show that the proposed MineHFPs algorithm for mining hierarchical frequent patterns is effective and efficient for complex queries.

Keywords: Object-oriented database · Mining frequent patterns · Inheritance hierarchy · Multiple data sources · Hierarchical frequent patterns

1 Introduction

Real world data are complex and good to be presented or modeled as objects. An object-oriented database is suitable for capturing more comprehensive and

This research was supported by the Natural Science and Engineering Research Council (NSERC) of Canada under an operating grant (OGP-0194134) and a University of Windsor grant.

A. Hameurlain et al. (Eds.): TLDKS XIII, LNCS 8420, pp. 158–186, 2014.
DOI: 10.1007/978-3-642-54426-2_6, © Springer-Verlag Berlin Heidelberg 2014

detailed complexity of real world data, such as different products on a Business to Customer (B2C) website, their histories, versions, price, images. Changes in contents or structure of a website may cause changes in the schema of the database that stores the web content. For example, a new product demonstrated on a B2C website which has its own specifications will need a different class object schema to store it [2,4,5,11]. Since the object-oriented database allows values of its attributes to be of complex types such as another database object, tables having attributes with type of new product do not need to change in structure, but only the new product object schema is also created. The attribute "new product" can be a set of new product classes whereby members of this set can take on any newly created product class schema as their type. An example schema representation for such a B2C web site is B2C(Webid:string, Products:set of products, NewProducts: set of products). With relational database system, values of all attributes of a table are single-valued such that the same B2C database schema above can be represented as B2C(Webid:string, product1:string, product2:string, newproduct1:string). If the web site gets new products, the B2C schema together with other schemas in the database need to be updated. The object oriented model presents a data structure for more clearly establishing complex relationships (e.g., superclass, subclass, part-of) between different data entities (e.g., classes and tables) so that mining of multiple tables, classes and databases on historical, derived and other data can be accomplished. The object schemas of the complex data types can be used to define version, histories, derived and other features of the products and new products so that when there are changes, only the relevant class structure needs to change in the object oriented database. Therefore, there is a great advantage in using an object-oriented database model to represent contents captured from web sites for comparative analysis as it presents a clear conceptual model that enables diverse, scalable mining of multiple databases which can still be implemented with the relational database management system (DBMS). Some recent work that also used a more realistic conceptual model such as the object oriented model being proposed in this paper to implement analysis of XML data (not multiple databases) include [21]. In an object-oriented database model, the same type of product (e.g., laptop) will be classified in the same class which inherits the properties (attributes) from its superclass (e.g., computer) and also has its own attributes. When a new product joins, a new class (e.g., pad) will be created for this type of product. For example, in a B2C website that sells computers and laptops, an object-oriented database to store the contents of this website has two classes, Computer and Laptop. The class "Computer" has the attributes "CPU", "RAM", "Hard_drive" while the class "Laptop" inherits the above three attributes of its superclass, "Computer" and also has its own attributes "Screen_size" and "Battery_life". If a new product, Pad which is a subtype of laptop comes to the website, a new class "Pad" will be created and it inherits the attributes of "Laptop" class and would also have its own attributes "3G_device" or "Touch_screen". An object-oriented database is a database management system in which information is represented in the form of encapsulated objects (possibly active) rather than static data values [8, 18, 22]. Due to the first normal form (1NF) requiring only single valued attributes, relational databases do not allow complex values, such as sets, lists, or other data structures. On the other hand, the attributes of an object-oriented database model can be a complex collection of types, such as, sets, lists, or some other data structure such as another class object. When implemented with an object-oriented database management system, the object oriented database model does not need additional tables to store the data represented in a collection type. In a relational database model, procedures (that is, transactions for manipulating the static data) must be maintained outside of the relational data model itself through mechanisms for querying and manipulating the data. However, in an object-oriented database model, these procedures can be considered as behaviors of the objects and can be maintained as methods of the classes.

2 Object-Oriented Database Schema

An object-oriented database model is represented by a set of classes connected by their class inheritance hierarchy through superclass and subclass relationships [9, 18]. An object-oriented database consists of a set of classes, C_i, with a class inheritance hierarchy H which is used to depict superclass and subclass relationships between classes in the object-oriented database and can be represented as a set of pairs of class and superclass in the form of (class, superclass). A superclass (e.g., Computer) of a class (e.g., Laptop), is a generalization of the class such that a class inherits all the attributes and methods of its superclass. Each class is defined as an ordered relation $C_i = (K, T, S, A, M, O)$, where K is the class identifier (e.g., computer id), T is the class type (e.g., Computer), S is the super type (superclass) of the class (e.g., Root), A is a set of attributes of the class (e.g., CPU, RAM, Hard_drive) [9]. M is a set of methods of the class (e.g., get_Number_of_Computer, get_Number_ofSalesof_Computer). O is a set of encapsulated instance objects (equivalent to tuples) of the class which have instances of the class attributes and methods (e.g., computers with specific CPU, RAM, Hard_drive and have instances of class methods). For example, a computer retail store object-oriented database consists of four classes (Root, computer, laptop, desktop), which are related through class inheritance hierarchy, H. H is specified in the format of all pairs of (class, superclass) relationships where H={(Root, Computer), (Laptop, Computer), (Desktop, Computer)}. Root class is the special class which exists in every database and is the only class with no superclass. Database schema for this database is provided as:

$C_1 = (K_1,$ Type, Super, {oid, CPU, RAM, Hard_drive, computer_name}, o_1, o_2, ..., o_n);
$C_2 = (K_2,$ Type, Super, {oid, Screen_size, Battery_life}, o_1, o_2, ..., o_n);
$C_3 = (K_3,$ Type, Super, {oid, Graphic}, o_1, o_2, ..., o_n);

Multiple object-oriented databases and classes can also be connected by their class inheritance hierarchy through superclass and subclass relationships with a multiple object-oriented database inheritance hierarchy MH which is used to

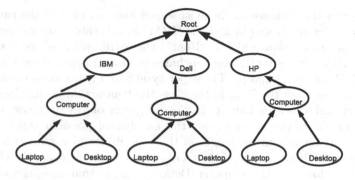

Fig. 1. The Multiple databases inheritance hierarchy Tree (MHTree) for 3 databases

depict superclass and subclass relationship between classes and object-oriented databases, and can be represented as a set of pairs of class and superclass in the form of (class, superclass). For example, the inheritance hierarchy in the multiple databases, MH for three computer object databases for IBM, Dell, and HP computers, has every database consisting of four classes (Root, computer, laptop, desktop). Note that when more than one database is integrated in the object schema, the Root class of each database becomes the name of the database. For example, the Root for the IBM database is now IBM since the integrated schema has only one global Root class. This MH is represented with the set of (subclass, superclass) relationships as follows. MH={(IBM, Root), (Dell, Root), (HP, Root), (Computer, IBM), (Computer, Dell), (Computer, HP), (Laptop, Computer), (Desktop, Computer)}.

Definition 2.01. *Tree structure of a multiple database class inheritance hierarchy (MHTree): is the tree structure representation of multiple databases inheritance hierarchy. For example, the MHTree for three object-oriented computer databases for IBM, Dell, and HP is shown in Fig. 1.* ∎

In a multiple database, inheritance hierarchy as shown in Fig. 1, there is a Root class. The database schema of the Root class is defined as Root(K, T, S, A, M, O). The Root class table is a transaction table which records the transactions on the classes in the multiple database. For example, sales transactions from different object databases can be recorded in the Root table. K in this Root table is the transaction id. T is the class type of the transaction (name of the database where the transaction comes from). S is the super type of the transaction. A is a set of attributes consisting of the set of super_type attributes of C_i class in T called $super_i$ (the number of $super_i$ depends on the levels of the hierarchy of C_i) and all attributes A of C_i. M is the set of class methods which are behaviors of the Root class, such as those for updating the Root table and mining patterns in the Root table. O is the set of instance objects of the transactions (one object stands for one transaction). For example, a sales transaction of a purchased laptop from IBM database recorded in the Root table (sales transaction table) has object id as an instance of K for transaction id (an integer number), class type T which indicates the database or the most senior ancestor class of the path where the transaction comes from (database is "IBM" in this case), super_type S of the T class of the transaction, which is "Root", the attributes A of the transaction class path (IBM/Computer/Laptop) includes two super types of the transaction class (IBM/Computer/Laptop). The super_type S of a transaction class consists of class hierarchy from the Root to the class the transaction is on. Thus, $super_1$ is Computer and $super_2$ is Laptop. In the computer object database, there are two levels of the hierarchy and thus, the number of possible $super_i$'s in the Root table corresponds to the length of the inheritance hierarchy. If a transaction record of the Root table concerns a desktop computer, then, the class path of this transaction is IBM/Computer/Desktop and in that case, the $super_1$ (is Computer) and $super_2$ (is Desktop). The attributes of all classes in the hierarchy make up the attributes of the Root class, A and in this case, they are: C_i, CPU, RAM, Hard_drive, Screen_size, battery_life and Graphic. In an object-oriented database model, the instantiated objects (instances) are referenced (retrieved) by following their object pointers. The relational model has the advantage of availability for its database management system for implementation purposes. Thus, we have chosen to have the current implementation of our object oriented mining (OOMining) model with the readily available relational DBMS. The relational database model also provides a clear visual conceptual representation of table schemas showing all attributes of a table (a relational table can be used to represent an object oriented class), including the information on the class inheritance hierarchy. However, converting the object-oriented database model into a relational model poses some challenges with regards to extensions needed for such operations as join operation between objects of different classes. In an object-oriented database model, there is no specific join operation, because the instantiated objects are referenced by the object pointers. We will provide the solution to address the problem of the object join operation in Sect. 4.1. We can also define the object-oriented database as a relational database represented with a set of tables (relations) as classes connected through foreign key relationships as inheritance hierarchy. The foreign keys in our object-oriented database model of object database ODB = a set of classes C_i including the Root class, where each C_i = (K, T, S, A, M, O) and Root(K, T, S, A, M, O) are realized through the inheritance hierarchy using the subclass and superclass relationships as defined in the class type T, supertype S attributes of each object class. For example, a relational database schema that represents the computer world object database and its Root class is shown below:

Computer (comp_id: string, type: string, super_type: string, cpu: string, ram: string, hard_drive: string);
Laptop (laptid: string, type: string, super_type: string, screen_size: string, battery_life: string);
Desktop (deskid: string, type: string, super_type: string, graphic: string);
Root (transactionid: integer, type: string, super_type: string, super1: string, super2: string, cpu: string, ram: string, hard_drive: string, screen_size: string, battery_life: string, graphic: string);

Table 1. Object table of computer class

Comp_id	Type	Super_type	CPU	RAM	Hard drive
Comp1	Laptop	Computer	2 GHz	2 GB	250 GB
Comp2	Laptop	Computer	2 GHz	2 GB	320 GB
Comp3	Laptop	Computer	3 GHz	4 GB	350 GB
Comp4	Desktop	Computer	3 GHz	4 GB	500 GB
Comp5	Desktop	Computer	3 GHz	4 GB	500 GB
Comp6	Desktop	Computer	3 GHz	4 GB	500 GB

Table 2. Object table of laptop class

Lap_id	Type	Super_type	Screen_size	Battery_life
Lapt1	Ideapad laptop	Laptop	15"	3 h
Lapt2	Ideapad laptop	Laptop	15"	3 h
Lapt3	Thinkpad laptop	Laptop	17"	3.5 h

In the above relational database schema, compid is the primary key of the computer table, laptid is the primary key of laptop table, deskid is the primary key of the desktop table, and transactionid is the primary key of Root table. All class tables have the composite foreign keys consisting of the two attributes "type" and "super_type" in each table. A computer object database for the respective classes of Computer, Laptop and Desktop is shown in Tables 1, 2, and 3.

Table 1 is the Computer class table that stores the specifications of computers and contains all instances of all computer types. Tables 2 and 3 store the specifications of laptops and desktops which inherit from the computer class. An example of the Root class table that records all computers purchased from different databases, such as IBM, Dell, or HP is shown in Table 4. Table 4 is a sales transactions table containing eight Root class instance objects where every object indicates one transaction of computer purchased from database specified in 'Type'. In the schema of the Root(K, T, S, A, M, O), "Oid" is the object id for each transaction. Of course, "oid" is an instance of K (transaction id), which is represented by an integer number. "Type" is the class type T of the transaction which indicates the database or the full inheritance path for the class involved in the transaction. For example, in transaction 1 of the Root table, it can be seen that the full class path for this transaction is "IBM/computer/laptop". Although in Type, the most senior ancestor class (IBM) in the path is recorded, the attributes of $super_1$ and $super_2$ will record the other classes of "computer" and "laptop" along this inheritance hierarchy of the transaction. Thus, "Types" are recorded as "IBM", "Dell", or "HP". The "Super_type" is the Root class in this case. The set of attributes (A) of the Root class, includes: (1) $super_1$ (computer) and $super_2$ (laptop or desktop) for the class of the transaction. In this example, computer class has subclasses laptop and desktop. There are two levels of the hierarchy and so there are 2 "super" attributes and for each $super_i$ attribute (e.g., at Computer class level), the domain (number of possible values)

Table 3. Object table of desktop class

Desk_id	Type	Super_type	Graphic
Desk1	Work station	Desktop	256M
Desk2	Work station	Desktop	256M
Desk3	Desktop	Desktop	512M

Table 4. An instance of the root class table

Oid	Type	Super type	Super1	Super2	CPU	RAM	Hard drive	Screen size	Battery life	Graphic
1	IBM	Root	Computer	Laptop	2 GHz	2 GB	250 GB	15"	3 h	
2	IBM	Root	Computer	Laptop	2 GHz	4 GB	320 GB	15"	3 h	
3	Dell	Root	Computer	Laptop	2 GHz	2 GB	350 GB	17"	3.5 h	
4	HP	Root	Computer	Desktop	3 GHz	4 GB	500 GB			256M
5	HP	Root	Computer	Desktop	3 GHz	4 GB	500 GB			256M
6	Dell	Root	Computer	Desktop	3 GHz	4 GB	500 GB			512M
7	IBM	Root	Computer	Laptop	2 GHz	2 GB	320 GB	15"	3 h	
8	HP	Root	Computer	Laptop	3 GHz	4 GB	350 GB	17"	3.5 h	

consists of its number of breadth-wise sibling classes itself included (e.g., it is one for class computer), while for $super_i$ class at the laptop level, the number of possible values is two consisting of the two sibling classes, laptop and desktop. Thus, with the example database, the Root class has two $super_i$ classes as $super_1$ (computer) and $super_2$ (laptop or desktop). If there are n levels of hierarchy, there will be $super_1, \ldots, super_n$. (2) CPU, RAM, Hard_driver, Screen_size, Battery_life, Graphic are all attributes of all the classes in the hierarchy consisting of computer, laptop, and desktop classes.

2.1 Frequent Pattern Mining in Object-Oriented Model

Frequent patterns are itemsets that appear in a data set with frequency (also called support) not less than a user-specified threshold (also called minimum support). Frequent pattern mining is the task of discovering frequent patterns from transactional databases. Frequent pattern mining is the essential step of association rule mining. Association rule is an implication of the form $X \rightarrow Y_i$, where X is a set of some items in the set of all items Y, and Y_i is a single item in Y that is not present in X. Frequent pattern mining in a single relational database table is used to find the itemsets whose frequencies over all transactions in the database table are no less than a user-specified threshold (also called minimum support). Therefore, frequent patterns in traditional database system consist of items or combination of items (itemsets). In an object database table, every object can be considered as one row (tuple) of a relational database table. The attributes of the object can be considered as object itemsets (patterns). Mining frequent patterns in object table is used to discover object attributes or combinations of object attributes that appear frequently in all objects of the class (or table) [15,20]. In Table 1, a computer class table has attributes "CPU", "RAM", "Hard_drive". The objects in Table 1 have attributes, such as < 2 GHz >, < 3 GHz >, < 2 GB >, < 4 GB >, or < 500 GB >. These attributes can be considered as itemsets. Based on Table 1 (sample of computer objects table), some frequent pattern mining queries that can be answered are:

Query 1: What are the most frequently used hardware components (CPU, RAM, hard drive) in IBM computer model products with a minimum support of 50 %? Query 1 can be answered by applying one of the frequent pattern mining algorithms, such as TidFP [12] on Table 1.

Query 2: What are the most frequently used hardware components (CPU, RAM, Screen size) in IBM laptop model products with a minimum support of 50 %?. Query 2 cannot be answered by applying TidFP algorithm on only computer class Table 1 or only on laptop class Table 1, because laptop IS-A-TYPE of computer and the computer class does not contain the specialization features of a laptop. Similarly, the laptop class alone does not contain the generalization features of a computer. Thus, to answer Query 2, there is need to involve the two tables, Tables 1 and 2. Tables 1 and 2 for classes computer and laptop need to be joined first, then we need to apply frequent pattern mining algorithms on the joined table. If we want to mine frequent patterns of the hardware specifications of computers that have been sold, we need to mine sales transaction table (shown in the Root Table 4). Assume that we want to answer the query such as:

Query 3: What are the most popular hardware component specifications (CPU, RAM, Hard_drive, screen size, battery life, and Graphics card) among the computer systems that have been sold with a minimum support of 50 %? If we apply TidFP algorithm on Table 4, we can only obtain the patterns in a format of transaction id list and itemset, <Tidlist, itemset >, <1,2,3,7, 2 GHz>, <4,5,6,8, 3 GHz>, <1,3,7, 2 GB>, <2,4,5,6,8, 4 GB>, <1,3,7, 2 GHz,2 GB>, and <4,5,6,8, 3 GHz,4 GB>. However, query 3 is not good enough to discover patterns in different hierarchies in an integrated multiple database table such as the Root Table 4. This table integrates information of hierarchy from multiple class tables in different databases using the object oriented data model. Therefore, we need the queries that can not only mine the frequent patterns, but also specify at which hierarchy level the pattern is frequent. For example, the queries such as:

Query 4: What are the most popular hardware component specifications (CPU, RAM, Hard_drive, screen size, battery life, and Graphics card) among the computer systems that have been sold by a particular company like Dell with a minimum support of 50 %?

Query 5: What are the most popular hardware component specifications (CPU, RAM, Hard_drive, screen size, and battery life) among a computer system subgroup such as laptops that are sold by a particular company like Dell with a minimum support of 50 %? To answer queries labeled as query 4 and query 5 (queries mining frequent patterns in transactional table), the algorithm is required to mine the attributes of computer, laptop, or desktop classes (computer, laptop, or desktop specifications) and also specify if the pattern is frequent at which hierarchy level.

Hierarchical Frequent Pattern. The TidFP algorithm [12] proposes a method that mines frequent patterns with transaction IDs to enable mining frequent patterns from more than one database table. With its technique, the resultant frequent patterns from more than one table are found by performing appropriate set operations of intersection, union and others on frequent patterns from different tables aided by common transaction ids from those tables as the tables were not pre-joined before mining. Thus, in TidFP, the frequent patterns are combinations of itemsets and their transaction id sets in the format of <Tidlist; itemsets>. Example queries such as Query 4 and Query 5 are looking for patterns that are frequent in different class hierarchy levels, and need to specify which hierarchy levels the pattern belongs to. Therefore, a new term, called hierarchical frequent pattern is defined.

Definition 2.11. *Frequent patterns specifying class hierarchy: Hierarchical Frequent Pattern, HFP: is represented in the format of <Tidlist; itemset; $class_i$ > and is used to indicate in which transactions and in which class hierarchy that a frequent pattern (itemset) appears. For example, a pattern <1,3,4; 2 GHz,2 G; laptop/computer/IBM> where 1,3,4 are transaction IDs(Tidlist), 2 GHz, 2 G are itemsets, and laptop/computer/IBM is the class hierarchy of the class starting from the class to the Root.* ■

Contributions and Outlines. The contributions of this paper are as follows.
1. In order to enable mining diverse data from more than one database and table (e.g., representing different B2C product web sites like CompUSA and BestBuy), we define an object-oriented class model where each database is represented by a set of object classes, their class inheritance hierarchy and the Root transaction class (Sect. 2). The inheritance hierarchy is specified as a set of type, supertype pairs. The database schema is defined as a set of object classes C_i, where $C_i =$ (K, T, S, A, M, O) for K its class id, T its class type, S its set of superclass types, A its set of attributes, M its set of methods and O its set of instance objects.
2. In Sect. 4, we define proposed techniques including: the method called Object-Oriented Join (OOJoin) which joins superclass table C_{super} and sub class table C_{sub} by selecting the tuples which have distinct object id, $C_{super}.K$ and $C_{sub}.K$ from the result of $C_{super} \bowtie C_{sub}$, that is, selected tuples with distinct object ids occur where $C_{super}.T = C_{sub}.T$ or $C_{super}.T = C_{sub}.S$.
3. We define the new term, hierarchical frequent pattern, HFP, formed as <Tidlist; Itemset; Hierarchy>, where Tidlist is a set of object ids drawn from the set of instances of K. Itemset is a set of class attributes drawn from the set A, and Hierarchy is a sequence of classes from Root to class, C_i (called in the pattern as $class_i$). Hierarchical frequent pattern specifies at which hierarchy level the pattern is frequent and is an extension of the TidFP's pattern <Tidlist; itemset>.
4. We propose an algorithm called MineHFPs that mines hierarchical frequent patterns to answer frequent pattern mining queries and specify at which hierarchy level the pattern is frequent by traversing the multiple database hierarchy tree (MHTree) with the 1-itemset candidate patterns and transaction IDs.
5. We extend the map-gen join used in TidFP algorithm to oomap-gen join for generating k-itemsets candidate patterns during the process of MineHFPs

algorithm to reduce the number of k-itemsets candidate patterns and avoid unnecessary intersecting of transaction ids by indexing the patterns using two position codes according to inheritance hierarchy, start position and end position and checking the position code before generating k-itemsets candidate patterns.

Section 3 has other related work, Sect. 5 has comparative analysis while conclusions and future work are presented in Sect. 6.

3   Other Related Work

Frequent pattern mining algorithms, such as Apriori [1,3,23] and FP-tree [16], can only mine frequent patterns from one single database table. They cannot discover frequent patterns from multiple tables and multiple data sources. And also they cannot discover patterns in different class hierarchies, as the inputs of these algorithms are simple transactional database tables with no class inheritance hierarchies. These frequent pattern mining algorithms such as Apriori and FR-tree, and TidFP algorithm take one database table as input. The database table contains a number of transactions (or tuples) to be mined. Each transaction contains one transaction id and the patterns (or attributes involved in the transaction such as tv, laptop, desktop). For example, in one transaction <1,a, b, c, d>, "1" represents transaction id while the items purchased by the transaction id are represented as "a", "b", "c", "d" represent patterns. The TidFP algorithm [12] mines frequent patterns first, generating frequent patterns with their transaction ids (called TidFp), then applying set operations on the TidFps to answer frequent pattern related queries across multiple database tables. The TidFp algorithm represents each frequent i-itemset as an m-attribute tuple of the form $< F_{i1} ; T_{1i1}, T_{2i1}, \ldots, T_{mi1} >$, where F_{i1} is the first frequent i-itemset, and T_{mi1} is the mth transaction id of the first frequent i-itemset. For example, given the minimum support of 50 % and a table with only 4 transactions with transaction ids $D_1 \ldots D_4$ where each transaction has a list of itemsets drawn from the list 1, 2, 3, 4, 5. The TidFp algorithm would find the list of frequent 1-itemsets as $F_1 = \{< 1, D_1, D_3 >, < 2, D_2, D_3, D_4 >, < 3, D_1, D_2, D_3 >, < 5, D_2, D_3, D_4 >\}$. This means that the 1-itemset 1 is frequent because it can be found in 2 database transactions D_1, and D_3. To find the 2-candidate itemsets, the algorithm would obtain the 2-itemset list by joining the same way the Apriori-gen would obtain those, but would now obtain the resulting transaction id list as the intersection of the transaction id list of the two joined patterns. Thus, a mapgen-join of the two 1-itemset patterns $< 1, D_1, D_3 >$ and $< 2, D_2, D_3, D_4 >$ will yield the resulting 2-itemset pattern $< 1, 2, D_3 >$. The TidFP algorithm does not mine frequent patterns in integrated object-oriented multiple databases with inheritance hierarchies, nor specify the hierarchy levels that patterns belong to and carries the extra overhead of using set operations to integrate discovered patterns from individual related tables. Existing work, such as in Mining Multi-level Association Rule [14], mining in distributed databases [6,7,13,17] replace the patterns by another pattern in higher or lower hierarchy

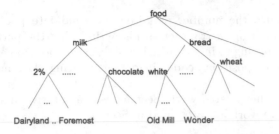

Fig. 2. The itemset concept hierarchy Tree

level and discover frequent patterns in different concept hierarchy level. However, these algorithms do not take object databases as inputs and do not consider the objects or object attributes as patterns.

For example (in Fig. 2), "2 % Foremost milk" is encoded as "112". Following the concept hierarchy, the digit "1" represents milk at level one, the second digit "1" represents 2 % milk at level 2, and the third digit "2" represents "Foremost" milk product at level three. For example, the transaction 1 in the transaction table is encoded as <1,111, 112, 211, 212>. In this transaction, the pattern "111" represents 2% Dairyland milk, the pattern "112" represents 2 % Foremost milk, the pattern "211" represents white OldMills bread, the pattern "212" represents white Wonder bread. Although the concept hierarchy provides some encoding of hierarchical semantic information about individual items (attributes), it is not the same as an object schema for representing the entire set of tables (classes) and the relationships between them. The OR-FP algorithm [19] takes object-oriented database as input and mines objects and attributes of objects as frequent patterns. However, it does not mine multiple object databases and does not specify at which hierarchy level patterns are frequent. The data in their object-oriented database is represented as: oi: class = $\{attribute_1, attribute_2, \ldots, attribute_n\}$. For example, parts of the sample data used by this system are: o1: Person = {Smith, Canada, 16000}; o2: Actor = {John, Canada, 12000}.

4   Mining Multiple Object-Oriented Databases

In this section, we define the object-oriented class model and a set of class methods in different classes. These class methods are able to integrate multiple data sources (by updating the Root class table), join object tables, and answer frequent pattern mining queries. The object-oriented database model consists of (1) a Root transactional class, Root (2) a set of object classes, $C_i \ldots C_n$, and (3) the inheritance hierarchy that defines the superclass-subclass relationships between the object classes, HTree. The structures of the three components are given as Algorithm 4.01 for the object database model.

Algorithm 4.01 *(The Object Database Model)*

OOModel()
begin
Root{
a set of transaction attributes A_i
//including super_type and all physical attributes of C_i
private void InsertTransactions;
private set MineRootFPs;
public set OOJoin;
}
Set of Classes $C_i \ldots C_n$ where for each C_i{
a set of physical attributes A_i
private set MineClassFPs;
}
Class Inheritance Hierarchy HTree in the form (subclass,superclass){
a set of (subclass,superclass) relationships
}
end

Class C_i has a set of physical attributes which are the properties of the class C_i. In the example of computer object database, physical attributes are "CPU", "RAM", and "Hard_drive" of a "Computer" class, or "Screen_size" and "Battery_life" of a "Laptop" class. Class Root has a set of transaction attributes. The transaction's attributes include a set of $super_i$ which consists of all the hierarchical super_types of the leaf class C_i and all physical attributes of classes C_i of the database. Private method InsertTransactions of Root class is used to insert transactions into the Root table and is only called in the class Root.

4.1 Object-Oriented Join (OOJoin)

To answer query 2 in Sect. 2.1, the computer class Table 1 and laptop class Table 2 need to be joined first. In the object database schema we defined in Algorithm 4.01, classes are connected by superclass and subclass relationships in the object-oriented database. Object-Oriented Join (OOJoin) is defined as a method which joins superclass and subclass tables on their type and super_type foreign keys. The main algorithm of OOJoin is shown as Algorithm 4.11.

Algorithm 4.11 *(OOJoin Algorithm)*

Algorithm OOJoin()
Input: Super class table C_{super}, sub class table C_{sub},
               superclass primary key K_1, the superclass foreign keys T_1 and S_1,
               subclass primary key K_2, the subclass foreign keys T_2 and S_2.
Output: A set of tuples of objects on Table T_d
Other Variables: Table T_c to hold result of cross product of
               two class tables, initialized as empty
               Table T_t for tuples of $C_{super}.T_1 = C_{sub}.T_2$ or

$C_{super}.T_1 = C_{sub}.S_2$, initialized as empty

List1: set of IDs of super class table, initialized as empty.

List2: set of IDs of sub class table, initialized as empty.

Begin

1.0 $T_c = C_{super} \times C_{sub}$. // cross product of tables

2.0 T_t = select from T_c where

$(C_{super}.T_1 = C_{sub}.T_2$ or $C_{super}.T_1 = C_{sub}.S_2)$

3.0 select a set of distinct tuples T_d from T_t;

3.1 insert the first tuple t_1 of T_t into T_d;

3.2 insert object id of superclass part in t_1 into $List_1$;

3.3 insert object id of subclass part in t_1 into $List_2$;

3.4 For each tuple t_x left in the T_t

    3.4.1 If (K_1 does not exist in $List_1$ and K_2 in t_1 does not exist in $List_2$)

        3.4.1.1 Insert t_x into T_d;

        3.4.1.2 Insert K_1 in t_x into $List_1$;

        3.4.1.2 Insert K_2 in t_x into $List_2$;

end

Description of the OOJoin Algorithm

The objective of the OOJoin algorithm is to join a class (e.g., laptop) with its superclass (e.g., computer) so that all inherited attributes of the class stored with the superclass can be obtained for queries of the class involving the inherited attributes as well. The OOJoin algorithm cascades from the class to the most senior ancestor class. Step 1.0 of the OOJoin algorithm finds the cross product of the super class and the sub class tables and stores the result in a temporary table T_c. The resulting tuples from the cross product operation contain all the attributes of the superclass (e.g., computer) and the subclass (e.g., laptop) including their primary and foreign keys. The subclass (laptop) keys consist of its primary key (K_2 such as laptid), its first foreign key which is the type for the subclass (T_2 such as laptop), and the second foreign key which is the super_type for the subclass (S_2 such as computer). Similarly, the joining superclass (computer) keys consist of computer class primary key (K_1 such as computer_id), first foreign key type (T_1 such as computer), and second foreign key for super_type (S_1 such as Root) are also in the attributes of this table, T_c. Step 2.0 of the OOJoin operation discards certain tuples from the result of the cross product operation according to the following conditions. For each tuple, the foreign key T_1 is compared with foreign key T_2. If T_1 matches T_2, or T_1 matches S_2 then the tuple will be kept, else the tuple will be discarded. Step 3.0 in the algorithm further prunes the list of tuples to keep only distinct tuples. The first tuple is always kept. Two lists are created, each list is a list of primary keys. The first list will be referred to as $List_1$ and it is used to store the K_1 primary keys and the other list is referred to as $List_2$ and it is used to store the K_2 primary keys. For each tuple, starting with the second tuple, we first check if K_1 of the current tuple is already in $List_1$. If it is, then this tuple will be discarded. Else, if K_1 is not already in $List_1$, then we check if K_2 is already in $List_2$. If it is, then the tuple will be discarded, else the tuple is kept.

Table 5. Result of OOJoin of Computer (C) class with Laptop(L) class

ID	Type	Super	CPU	RAM	Hard drive	Comp name	ID	Type	Super	Screen size	Battery life
Comp1	L	C	2 GHz	2G	250 G	I. laptop	Lapt1	Ideapad	L	15"	3 h
Comp2	L	C	2 GHz	2G	320G	I. laptop	Lapt2	Ideapad	L	15"	3 h
Comp3	L	C	3 GHz	4G	350 G	T. laptop	Lapt3	Thinkpad	L	17"	3.5 h

For example, OOJoin operation of Tables 1 and 2 will result in Table 5. The three tuples (comp1, comp2, comp3) of Table 5 join the computer class with laptop class to select all laptop instances with their inherited attributes specified in the join operation to select joined tuple from the cross product if $class_1$.type $= class_2$.supertype or $class_1$.supertype $= class_2$.supertype. With this join, if $class_1$.type $= class_2$.supertype or Table 2. supertype, then the two tuples of Tables 1 and 2 are joined. For example, for tuple comp1, comp1.Type = Laptop in Table 1 and lapt1.supertype = Laptop in Table 2 and these two tuples are joined to obtain tuple comp1 of Table 5. Other results are obtained in similar fashion.

4.2 Mining Frequent Patterns in One Class

The MineClassFPs algorithm is used to mine frequent patterns of any class. This it does by using the OOJoin algorithm to obtain all inherited attributes and methods of the class from its superclasses before it applies either the TidFp algorithm for mining the frequent patterns at different hierarchy levels of the inheritance hierarchy. As shown in the class model, every class C_i has a private method MineClassFPs which mines frequent patterns in the class and outputs a set of class attributes as frequent patterns. The algorithm for MineClassFPs is provided as Algorithm 4.21.

Algorithm 4.21 *(MineClassFPs Algorithm)*

Algorithm MineClassFPs()
Input: class table C to be mined, super class tables (CS_i) of class C
             where CS_k is superclass of CS_{k-1}, minimum support $s\%$
Other Variables: Joined class table T
Output: A set of frequent patterns FPs.
Begin
1.0 // Call JoinClasses (C, CS_i) to join classes as in step 1.x below
        1.1 T = C;
        1.2 if ($CS_i \neq$ NULL) // C has super classes.
                1.2.1 For each superclass table CS_i
                        1.2.1.1 T = OOJoin(CS_i , T); // call OOJoin to join
                        //subclass and superclass
2.0 TidFP(T, $s\%$); // Call TidFP on Joined table T
end

Description of the MineClassFPs Method

Step 1.0 of the algorithm joins the class and all super classes using the OOJoin algorithm. Step 2.0 applies TidFP algorithm which takes resulting table from Step 1.0 and the minimum support to mine the frequent patterns. This private method MineClassFPs can answer queries such as query 2 in Sect. 2.1.

4.3   Mining Hierarchical Frequent Patterns in the Root Class

As shown in the class model (Algorithm 4.01), the Root class has a private method MineRootFPs (as given in the formal algorithm 4.51). This method mines frequent patterns in the class and outputs a set of frequent patterns specifying the levels of the inheritance hierarchy. The hierarchical frequent patterns are mined from a Root table of transactions on classes (tables) in the inheritance hierarchy such as Table 4. The inheritance hierarchies exist in the transaction in the Root table. The algorithm for mining hierarchical frequent patterns first creates multiple database inheritance hierarchy tree (MHTree), such as Fig. 1. Then, the transaction ids of the Root table are stored in the MHTree node according to the inheritance hierarchy of each transaction in the Root table. Then, the algorithm traverses the MHTree through the linkage table to access every node and intersect the transaction ids in every node with transaction ids of 1-itemset candidate patterns to obtain 1-itemset frequent patterns with the hierarchy information. A modified version of map-gen join algorithm in the TidFP algorithm is used to generate 2-itemset candidate pattern, and it then traverses the MHTree to obtain 2-itemset frequent patterns. Finally, it obtains the n-itemset frequent patterns. The process of mining the hierarchical FP from the Root table is given in algorithm 4.51 which starts by obtaining the multiple inheritance tree (MHTree) and this calls the OOJoin algorithm to join each such subclass (e.g., laptop) with its superclass (e.g., computer) so that all inherited attributes of the class stored with the superclass can be obtained. The MineClassFP algorithm 4.21 for mining frequent patterns of any class also uses the OOjoin algorithm to obtain inherited attributes of all superclasses of this class. To mine only class FP, the algorithm would use OOJoin to obtain the full class information before applying the TidFP algorithm to obtain the class FPs with Tids. To mine the Root FPs, the OOJoin is used to create the MHTree before calling the MineHFPs with inheritance hierarchy information about each transaction to generate hierarchical FPs where each iteration involves oomap gen join of frequent F_k with itself.

4.4   Position Coding Method

In the PLWAPLong algorithm [10], two position codes, start position and end position (two integer numbers) are assigned to every node of the tree to distinguish the position of the nodes in the tree. Position codes are assigned by pre-order traversal of the tree (in the order visit node, left subtree and right subtree) and starting with the root node of the tree having the start position code of 0. The idea of position coding method can be used to represent the

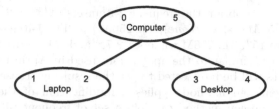

Fig. 3. The position code assigned HTree

level of inheritance hierarchy. As shown in Fig. 1, multiple database inheritance hierarchy can be represented in a tree structure called MHTree, inheritance hierarchy in one database can also be represented in a tree structure called HTree. Position coding method can be used to assign two position codes, start position and end position to each node of the HTree/MHTree by pre-order traversal in order to represent the levels of inheritance hierarchy. The position code assigned HTree of the example computer database is shown in Fig. 3.

In Fig. 3, two position codes are assigned to every node of the HTree. The start position of the Computer class (root parent class) is "0" which is less than the start position of the laptop class (child class) and the desktop class (child class). The end position of the computer class is "5" which is greater than the start positions of the child classes. The laptop class and the desktop class are the sibling classes. The start position and end position of one sibling class are both smaller than those of the other sibling class or the start position and end position of one sibling class are both greater than those of the other sibling class.

The oomap-gen Join Method
Like in the Apriori-gen join, the purpose of the oomap-gen is to obtain the extended (i + 1)-itemsets from the frequent i-itemsets (F_i) by joining F_i with itself oomap-gen fashion. The map-gen join method used in the TidFP algorithm avoids multiple database scanning by intersecting transaction id lists of two patterns being joined to get the resulting transaction id list. The resulting itemset is obtained as the union of the two joined itemsets. However, map-gen join still suffers from large candidate generation and intersecting transaction id lists of every candidate patterns and unable to apply to object hierarchy. When the number of transactions is large, intersecting transaction id lists is an expensive process. Figure 4 provides an example application of the map-gen join of patterns from the example sales transaction table (Root table) shown in Table 4. The patterns in map-gen join are in the slightly reordered format (where Tidlist comes before the itemset list) of <Tidlist; itemset>. In that Fig. 4, it can be seen that the computer feature attribute 1-itemset of <15">, <3 h>, and <256M> are all 1-frequent items where <15">, and <3h>, are both frequent in the Root table transactions with ids 1, 2 and 7. However, the 1-itemset <256M> is frequent in Root table transaction ids 4 and 5. The goal of the map-gen join of these three frequent 1-itemsets $< (1, 2, 7); (15'') >$, $< (1, 2, 7); (3 \, h) >$, $< (4, 5); (256M) >$

with themselves, is to obtain the frequent 2-itemsets of $< (1, 2, 7); (15'', 3h) >$, $< (None); (15'', 256M) >$, $< (None); (3h, 256M) >$. The 1-itemsets in the map-gen join above are $15''$, 3 h, 256M while <1,2,7> and <4,5> are the transaction id lists. This step is similar to the ap-gen join used in Apriori algorithm. The transaction id lists will be intersected to get the resulting transaction id list.

The oomap-gen join method applies a modification of the map-gen join function. The oomap-gen method can join a set of frequent i-itemsets F_i with itself, where itemsets are from an object oriented class inheritance hierarchy, to obtain the candidate (i+1)-itemsets. Thus, the candidate (i+1)-itemsets C_{i+1} is obtained from the frequent i-itemsets for $i \geq 1$, by joining frequent i-itemsets F_i with itself oomapgen way such that $C_{i+1} = F_i \bowtie F_i$. To join oomapgen way, for each pair of itemsets M and $P \in F_i$ where each F_i itemset is in the format "< transaction id list, itemset, (class start position code, class end position code) >", the following three conditions have to be satisfied: M joins with P to get itemset $M \cup P$ if the following conditions are satisfied.
(a) itemset M comes before itemset P in F_i,
(b) the first i-1 items in M and P (excluding just the last item) are the same,
(c) the transaction id list of the new itemset $M \cup P$ represented as $Tid_{M \cup P}$ is obtained as the intersection of the Tid lists of the two joined i-itemsets M and P and thus, $Tid_{M \cup P} = Tid_M \cap Tid_P$.
(d) To speed up processing, ignore non-joinable patterns by applying the oomap pattern joinable rule which states that only patterns belonging to the same class or classes with ancestor-descendant relationships determined using the start and end position codes of the patterns are joinable.

Definition 4.41. *Ancestor-Descendant Nodes (a,b): Node a of a tree is an ancestor of node b of the tree iff the start position code of node a is less than the start position code of node b, but the end position code of node a is greater than the end position code of node b. For example, in the Htree of Fig. 3, the node Computer with (start,end)codes of (0, 5) is an ancestor of the node Laptop with codes (1, 2).* ■

Definition 4.42. *Sibling Nodes (a,b): Node a of a tree is a sibling of node b of the tree iff both the start and end position codes of node a are either less than or greater than the start and end codes of node b. For example, in the Htree of Fig. 3, the node Laptop node with (start,end)codes of (1, 2) is a sibling of the node Desktop with codes (3, 4).* ■

Definition 4.43. *The oomap pattern Join Rule: If two patterns belong to the same class or belong to two different classes but have an ascendant-descendant relationship as can be determined with Ancestor-Descendant Node definition, they can be joined. If two patterns belong to different classes which have a non ascendant-descendant relationship, they cannot be joined. For example, using this rule for $< (1, 2, 7)(15'') > (1, 2)$ oomap-gen join $< (4, 5)(15'') > (3, 4)$ patterns will yield no join since from the ancestor-descendant rule the start and end position codes of the first pattern are all less than those of the second pattern showing that they are not from joinable classes.* ■

<(1,2,7);(15">   map <(1,2,7);(15")>
<(1,2,7);(3hrs)> -gen <(1,2,7);(3hrs)>   =
<(4,5);(256M)>   Join <(4,5);(256M)>

<(1,2,7);(15",3hrs)>
<(None);(15",256M>
<(None);(3hrs,256M>

Fig. 4. The map-gen join

<(1,2,7);(15")>(1,2)   oomap-gen   <(1,2,7);(15")>(1,2)
<(1,2,7);(3hrs)>(1,2)   join   <(1,2,7);(3hrs)>(1,2)
<(4,5);(256M)>(3,4)   <(4,5);(256M)>(3,4)

=

<(1,2,7);(15",3hrs)>(1,2)

Fig. 5. The OOmap-gen join

From Fig. 4, it can be seen that applying map-gen join on three 1-itemset patterns will result in three 2-itemset candidate patterns. The transaction id lists of newly generated 2-itemset patterns <15", 256M> and <3h, 256M> are None, because the patterns <15"> and <3h> belong to the laptop class, but the pattern <256M> belongs to the desktop class, they cannot appear in the same transaction in the sales transaction table. Therefore, the position coding method introduced in the previous section will be used to reduce the candidate pattern generation. With the position codes involved, patterns will be checked for their inheritance hierarchy relationships before generating the new candidate patterns. The start position and end position can be used to distinguish the ascendant-descendant or sibling relationships of classes. As given in the oomap-gen pattern join rule, if two patterns belong to the same class or belong to two different classes but have an ascendant-descendant relationship as can be determined with Ancestor-Descendant Node definition, they can be joined. If two patterns belong to different classes which have a non ascendant-descendant relationship, they cannot be joined. Figure 5 shows the result of the oomap-gen join where the start and end position codes may be used to more quickly identify the non-joinable patterns such as $< (None); (15'', 256M >$ that appeared in the result of the map-gen join of Fig. 4 and to exclude them in the result of the oomap-gen join as shown in Fig. 5.

From Fig. 5, it can be seen that patterns are in the format of <Tidlist; itemset>(start, end).

4.5   The MineRootFPs Method

The main algorithm of MineRootFPs method is used for answering comparative queries involving transactions of many classes in the object database which can also include an integration of multiple databases such as computers from several vendors like IBM, Dell, HP as shown in the multiple inheritance hierarchy of Fig. 6. The formal algorithm for MineRootFPs(MH, $s\%$, Root) is given as Algorithm 4.51.

Algorithm 4.51 *(MineRootFPs Algorithm)*

Algorithm MineRootFPs()
Input: multiple database inheritance hierarchy MH,
        Root table, minimum support $s\%$
Other variables: multiple database inheritance hierarchy Tree MHTree,
        TMHTree, //Transaction ids stored MHTree
        LTMHTree //Linkage built TMHTree,
        set of k-itemset frequent pattern F_k;
        set of k-itemset candidate pattern C_k;
Output: hierarchical frequent patterns HFPs in the format of
        <Tidlist; itemsets; $class_i$ >.
Begin
1.0 CreateMHTree(MH);
//create multiple database inheritance hierarchy tree, MHTree
2.0 StoreTidMHTree(MHTree, Root);
//store transaction ids into MHTree and Obtain TMHtree
3.0 GenOneCand(Root); //generate 1-itemset candidate patterns
4.0 BuildLinkage(TMHTree); //build linkage of TMHTree and obtain TMHTree
5.0 MineHFPs(LTMHTree, C_k, $s\%$)
5.1 C_k = 1-itemset candidate patterns
5.2 F_k = CheckMinS(MHTree, C_k, $s\%$);
5.3 if (F_k is not empty)
      5.3.1 C_{k+1} = oomap-gen-join(F_k);
      5.3.2 $k = k + 1$
      5.3.2 go to step 5.2
End

Description of MineRootFPs Algorithm of the Root Class

Step 1.0 is creates a multiple database inheritance hierarchy tree (MHTree) as shown in Fig. 6. Step 2.0 scans the entire transaction Table 4 and stores the transaction ids into the nodes (class) of the MHTree to create a MHTree with transaction ids stored which is called TMHTree. For example, from Fig. 6, it can be seen that transactions 1, 2, 7 of the Root transaction Table 4 are on IBM laptop computers. Concurrently with steps 2.0, step 3.0, it generates the 1-itemset candidate patterns in the format of <Tidlist; itemset>(start, end). This step is similar to the step of generating 1-itemset candidate patterns in the TidFP algorithm. Step 4.0 is building the header linkage to TMHTree to create a LTMHTree, so that nodes of the tree can be easily accessed. The header linkage

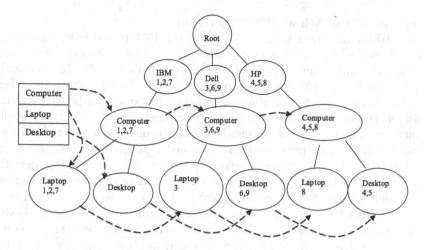

Fig. 6. The LTMHTree:linkage tree multiple inheritance tree

is built such that every unique class in a database (e.g., Computer, Laptop, Desktop) has an entry and using the preorder traversal, all similar classes across multiple databases are linked in a queue. In the computer object database, there are three object tables, "Computer", "Laptop", and "Desktop". Therefore, there will be three entries in the link header table. It builds linkage queue for each entry. Finally, it uses pre-order traversal (visit node, visit left subtree, visit right subtree) to access every node of the tree and store the node into the appropriate queue. An LTMHTree with transaction ids stored and linkage built is shown in Fig. 6. Step 4.0 is for mining the hierarchical frequent patterns in the Root table by calling the MineHFPs algorithm.

Algorithm 4.52 *(MineHFPs Algorithm)*

Algorithm MineHFPs()
Input: linkage built, transaction ids multiple database
        inheritance hierarchy LTMHTree, minimum support $s\%$,
        a set of 1-itemset candidate pattern C_1, in the format of <Tidlist, itemset>
Output: a set of hierarchical frequent patterns F_k
        in the format of <Tidlist, itemsets, $class_i$ >
Other variable: a set of candidate patterns C_k
Begin
1.0 $C_k = C_1$
2.0 F_k = CheckSupp(LTMHTree, C_k, $s\%$);
3.0 if (F_k is not empty)
3.1 C_{k+1} = oomap-gen-join(F_k);
3.2 k = k+1
3.3 go to step 2.0
end

Description of the MineHFPs Algorithm

The MineHFPs algorithm takes as an input the LTMHTree (with transaction IDs stored and linkage built), a set of 1-itemset candidate patterns with transaction IDs, and a minimum support value $s\%$. The algorithm MineHFPs calls the algorithm CheckSupp which uses every 1-itemset candidate pattern to traverse LTMHTree in order to check the support of each 1-itemset candidate pattern. If the support is greater than or equal to the minimum support of $s\%$ at any level in the hierarchy, then the 1-itemset candidate pattern counts as a 1-itemset frequent pattern. If 1-itemset frequent pattern(s) already exists, it uses oomap-gen-join algorithm to generate 2-itemset candidate patterns. The CheckSupp algorithm is utilized to check the support level of the newly generated 2-itemset candidate patterns and to generate 2-itemset frequent pattern(s) if the support level is sufficient. If 2-itemset frequent patterns exist, it uses algorithm oomap-gen-join to generate 3-itemset candidate patterns. By the same process, k-itemset frequent patterns can be generated. The CheckSupp algorithm is given as Algorithm 4.53.

Algorithm 4.53 *(CheckSupp Algorithm)*

Algorithm CheckSupp()
Algorithm CheckSupp(LTMHTree, Ck, $s\%$);
Input: MHTree, k-itemset candidate pattern with transaction IDs C_k,
    in the format of $<$Tidlist, itemsets, $class_i >$, k $= 1$; initially, minimum support $s\%$.
Output: Frequent k-itemsets F_k, in the format of $<$Tidlist, itemsets, $class_i >$.
Other variables: intersected transaction id list intersectTidlist, unioned Transaction
    id list UTidlist, Pointer nodePtr, Frequent pattern f, Boolean Flag=false,
    linkage queue of LTMHTree q_i, Hierarchy of every node $class_i$
Begin // Check supports of generated patterns
1.0 For each element c_x in C_k do
    1.1 Flag = false;
    1.2 For each queue qi do
        1.2.1 For each element e_{ij} in the queue qi do
        1.2.1.1 intersectTidlist = $c_x.Tidlist \cap e_{ij}.Tidlist$;
        1.2.1.2 if((number of IDs in intersectTidlist)/(number of
        IDs in $e_{ij}.Tidlist$) $>= s\%$)
            1.2.1.2.1 f = c_x; 1.2.1.2.2 insert f into F_k;
            1.2.1.2.3 f = f append $e_{ij}.class_i$; 1.2.1.2.4 Flag = true;
        1.2.1.3 UTidlist = UTidlist $\cup e_{ij}.Tidlist$;
        1.2.2 intersectTidlist = $c_x.Tidlist \cap$ UTidlist;
        1.2.3 if((number of IDs in intersectTidlist)/(number of IDs in UTidlist) $>= s\%$)
            1.2.3.1 insert f into F_k; 1.2.3.2 f = c_x concatenate $e_{ij}.class_i$;
            1.2.3.3 Flag = true;
        1.3 if(Flag = true) Insert c_x into F_k;
end

Application of the CheckSupp Algorithm

To serve as an example, the MineHFPs algorithm uses the inputs LTMHTree (Fig. 6), two 1-itemset candidate patterns: $<$1, 3, 7, 2 GHz$>$ (0,5) and $<$1, 3, 7, 2 G$>$ (0,5), and a minimum support value of 50 %. Step 1.0 and 2.0 of the MineHFPs algorithm use the transaction id list (Tidlist) of every 1-itemset candidate pattern to intersect the Tidlist of every node in every linkage queue of the LTMHTree in order to discover the 1-itemset frequent patterns. The MineHFPs algorithm starts from the first 1-itemset candidate pattern <1, 3, 7, 2 GHz>. The Tidlist of the candidate pattern <1, 3, 7, 2 GHz> is <1, 3, 7>. The first node of linkage queue of "Computer <1, 2, 7>" is <1, 2, 7> (according to Fig. 3.13). Intersecting <1, 3, 7, 2> and <1, 2, 7> obtains <1, 7>. There are two transaction ids in <1, 7>. The number of ids in <1, 2, 7> is 3. The frequency is 2/3 which is greater than 50 %. Hierarchy of node "Computer <1, 2, 7>" is node "computer/IBM". Therefore, we obtain the hierarchical frequent pattern <1, 7, 2GHz, computer/IBM>. We also insert the candidate pattern <1, 3, 7, 2 GHz> into frequent pattern set F_1. In the same way of processing the candidate pattern <1, 3, 7, 2 GHz> and node "Computer <3,6,9>" is intersected, and pattern <1, 3, 7, 2 GHz> and node "Computer <4,5,8>" is intersected. We find out that pattern <1, 3, 7, 2 GHz> is not frequent at node "Computer <3,6,9>" nor at node "Computer <4,5,8>". We also need to union the Tidlists of all three nodes in the "Computer" linkage queue. Union of Tidlists <1,2,7>, <3,6,9>, and <4,5,8> is <1, 2, 7, 3, 6, 9, 4, 5, 8>. Intersecting Tidlist of pattern <1, 3, 7, 2, 2 GHz> and <1, 2, 7, 3, 6, 9, 4, 5, 8> is <1, 3, 7, 2>. The frequency is 4/9 which is less than the minimum support of 50 %. Therefore the pattern <1, 3, 7, 2, 2 GHz> is not frequent at the hierarchy "Computer". The Tidlist of candidate pattern <1, 3, 7, 2, 2 GHz> will intersect Tidlist of nodes in "Laptop" linkage queue and "Desktop" linkage queue. The 1-itemset candidate pattern <1, 3, 7, 2 G> will be processed by the same procedure as above and will obtain patterns as: <1, 7, 2 G, computer/IBM>, <1, 2, 2 G, laptop/computer/IBM>. Patterns <1, 3, 7, 2 GHz>(0,5) and <1, 3, 7, 2 G> (0,5) will use oomap-gen join to generate 2-itemset candidate pattern <1,3,7, 2 GHz, 2 G>(0,5). This 2-itemset pattern will serve as inputs to the CheckSupp algorithm and 2-itemset frequent patterns are generated. Then the 2-itemset frequent patterns will be used to generate 3-itemset candidate patterns by oomap-gen join. By the same process we obtain all k-itemsets hierarchical frequent patterns, until there are no frequent patterns generated.

5   Implementation and Experimentation

One of the most important contributions of the paper is proposing an object oriented model for representing and mining data from multiple databases while maintaining the class inheritance hierarchy for purposes of answering more complex, historical, derived queries across such integrated multiple database data. The experiments below serve to show both the effectiveness of the proposed algorithms in performing such object oriented mining while remaining reasonably efficient in comparison with existing system such as the TidFP that cannot handle all tasks that can be handled by the proposed approach. To test the performance of our proposed method for mining hierarchical frequent patterns in table Root (transaction table), we use the IBM quest synthetic data generator to generate three datasets for the three databases. There are three datasets (class object table C_i) in every database, the first one represents the "Computer" objects table, the second for the "Laptop" objects table, and the third for the "Desktop" objects table.

5.1   Generating the Class Table C_i

The IBM quest synthetic data generator generates integer numbers to represent patterns (attributes of objects in the case of object-oriented databases). If we specify the number of items, $\|N\|$, as "15", it means that the patterns will be represented by the integer numbers from "1" to "15". When we use the IBM quest synthetic data generator to generate the dataset which represents the Computer class table, we specify $\|N\|$ as "15". This means that the integer numbers from "1" to "15" will represent the patterns of the Computer class table. When we generate the dataset for the Laptop class table, we specify $\|N\|$ as "60". However, the integer numbers from "1" to "15" have already been used to represent the patterns for the Computer class table. We need to eliminate the numbers "1" to "15" so that the dataset generated will only contain the integer numbers from "16" to "60". Therefore, the integer number from "16" to "60" will be used to represent patterns for the Laptop class table. When we generate the dataset for the Desktop class table, we specify $\|N\|$ as "120". Since the numbers from "1" to "15" have already been used to represent the patterns for the Computer class table and the integer numbers from "15" to "60" have already been used to represent the patterns for the Laptop class table, we need to eliminate the numbers from "1" to "60" so that the dataset generated will only contain the integer numbers from "60" to "120". Therefore, the integer numbers from "60" to "120" will be used to represent patterns for the Desktop class table. Each transaction of the dataset represents one instantiated object. The transaction id of a transaction record will represent the object id of one instantiated object and a set of items in a transaction record will represent a set of object attributes in one instantiated object. We generate three datasets (Computer, Laptop, Desktop) for each of the three databases (IBM, Dell, and HP). We use an integer number to represent a particular database (the database name) and an integer number to represent a particular class object table. For example, "1" represents the "IBM" database, "2" represents the "Dell" database, "3" represents the "HP" database, "4" represents the "Computer" class, "5" represents the "Laptop" class, and "6" represents the "Desktop" class. The "Computer" class is inherited by the "Laptop" and the "Desktop" class. As discussed in Sect. 1.2, the database schema of C_i is C_i (K, T, S, A, M, O). T is the type and S is the super type. The dataset that stands for the "Computer" class will be assigned a number "4" as S (super_type), and randomly assigned "5" or "6" as T (type). With regards to the "Laptop" class, S(super_type) will be assigned as "5", and T (type) will be randomly assigned as "5","7" or "8" (which represent different subclasses of the "Laptop" class). With regards to the dataset that stands for the "Desktop" class, S(super_type) will be assigned as the number "6", and T (type) will be randomly assigned as "6","9" or "10" (which represent different subclasses of the "Desktop" class).

5.2   Generate the Root Table

The Root table is a transaction table that has transaction id K as a primary key, T and S as foreign keys (which represent type and super type of the transactions in Root table). K is the transaction id which is an integer number from 1 to $\|D\|$ sequentially. $\|D\|$ is the number of transactions in the Root table. Type, T, is used to represent the name of the database where the transactions come from. We randomly generate an integer number among "1", "2", "3" for type, T, for every transaction to represent the name of a database (such as IBM, Dell, and HP). Then we apply OOJoin algorithm to join all class tables C_i in every database to obtain an object joined table. Finally, randomly select the objects from object joined table in each database to fill in the attributes A in the Root Table.

5.3   Performance Comparison

The proposed algorithm MineHFPs is compared with the TidFP algorithm with respect to CPU execution time and memory usage because it is the algorithm that is closest to being able to answer the types of mining queries involving multiple tables and databases which the proposed algorithm and model is designed for. The most important contribution of work is providing a model that can mine multiple database tables and answer such complex queries involving history and derived data. It should also be mentioned that while the proposed approach mines frequent patterns in integrated or joined tables (classes), the TidFP mines FPs from individual tables and integrates the FPs to answer the query through relevant set operations. Thus, this could also be a reason for slower execution time for the TidFP in comparison with the MineHFP in some of the reported experiments.

MineHFP and TidFP are both implemented in C++ with the same data structures and can run on both windows and UNIX platforms. In a UNIX environment, the programs are compiled with "g++ filename" and executed with "a.out". The class object table C_i, inheritance hierarchy H, and multiple database inheritance hierarchy MH are all stored in text files. If we separate the integrated Root table by class hierarchy, the TidFP algorithm can also be applied to each separated part to answer those queries. For example, using the TidFP algorithm to answer "Query 4: What are the most popular hardware component specifications (CPU, RAM, Hard-drive, screen size, and battery life) among a computer system subgroup such as laptops and sold by a particular company like Dell (with a minimum support of 50 %)?". We will select transactions having type as "3" (transaction comes from Dell database), and also have super1 "4" and super2 "5" to represent "Computer" and "Laptop", respectively. In this section, we compare the performance of our proposed algorithm MineHFPs and the TidFP algorithm. Both the CPU execution times and the memory usages are measured for each algorithm. The MineHFPs algorithm performance measures include the tasks of creating the MHTree, storing transaction ids in the MHTree, generating 1-itemset candidate patterns, building linkage, and executing the MineHFPs algorithm. We generate the Root tables of size 125K, 250 K, 500K, and 1M. The characteristics of the generated datasets are described in Table 6. Table 7 describes the execution times for the MineHFPs and the TidFP algorithm on 125K dataset with low minimum support (20 %,

Table 6. The characteristics of the generated dataset

Root	Computer	Laptop	Desktop
Table	Class	Class	Class
125K	C7.S4.N20.D125K	C15.S4.N60.D63K	C25.S4.N120.D62K
250K	C7.S54.N20.D250K	C15.S4.N60.D125K	C25.S4.N20.D125K
500K	C7.S4.N20.D500K	C15.S4.N60.D250K	C25.S4.N20.D250K
1M	C7.S4.N20.D1000K	C15.S4.N60.D500K	C25.S4.N20.D500K

Table 7. CPU execution time on 125K dataset with varying minimum support

Algorithms (minimum Support)	Execution times (secs) at varying minimum supports				
	20 %	10 %	9 %	8 %	7 %
MineHFPs	290	4186	6356	9606	17785
TidFP	279	12327	23083	40097	74046

Table 8. Memory usage on 100K dataset with varying minimum support

Algorithms (minimum support)	Memory usage (in MB) at varying minimum supports				
	20 %	10 %	9 %	8 %	7 %
MineHFPs	62	430	590	774	1070
TidFP	26	158	214	266	350

10 %, 9 %, 8 %, and 7 %). Table 8 describes the memory usage of the MineHFPs and the TidFP algorithm on 125K dataset with low minimum support (20 %, 10 %, 9 %, 8 %, and 7 %). Table 9 gives the execution time of the MineHFPs and the TidFP algorithm on 250K dataset with low minimum support (20 %, 10 %, 9 %, 8 %, and 7 %). Table 10 gives the memory usage of the MineHFPs and the TidFP algorithm on 250K dataset with low minimum support (20 %, 10 %, 9 %, 8 %, and 7 %). Table 11 is the execution time of the MineHFPs and the TidFP algorithm on 500K dataset with low minimum support (20 %, 10 %, 9 %, 8 %, and 7 %). Table 12 is the memory usage of the MineHFPs and the TidFP algorithm on 500K dataset with the low minimum support (20 %, 10 %, 9 %, 8 %, and 7 %).

From Tables 9, 10, 11, we can see that the MineHFPs algorithm outperforms the TidFP at the low minimum support thresholds. The MineHFPs algorithm is approximately 3.5 times faster than the TidFP algorithm for a 125K dataset, 3.9 times faster for a 250K dataset, and 4.4 times faster for a 500K dataset when the minimum support is lower than 20 %. As the size of the dataset is increased, the performance margin between the MineHFPs and the TidFP algorithm increases in favor of the MineHFPs algorithm. From these tables, we can see that the MineHFPs algorithm has greater memory usage compared with the TidFP algorithm. The memory usage of the MineHFPs algorithm is approximately 2.8 times, 2.5 times, and 2.6 times greater than the TidFp algorithm for respective dataset sizes of 125K, 250K, and 500K (at the minimum supports

Table 9. CPU execution time on 250K dataset with varying minimum support

Algorithms (minimum support)	Runtime (in Seconds) at different supports)				
	20 %	10 %	9 %	8 %	7 %
MineHFPs	584	8321	12382	19241	35281
TidFP	577	24008	43584	74432	Crashed

Table 10. Memory usage on 250K dataset with varying minimum support

Algorithms (minimum support)	Memory usage (in MB) at varying minimum supports				
	20 %	10 %	9 %	8 %	7 %
MineHFPs	114	814	1098	1145	2001
TidFP	46	282	422	490	Crashed

Table 11. CPU execution time on 500K dataset with varying minimum support

Algorithms (minimum support)	Runtime (in Seconds) at different supports				
	20 %	10 %	9 %	8 %	7 %
MineHFPs	1180	16233	24679	37514	68143
TidFP	1150	48077	85027	Crashed	Crashed

Table 12. Memory usage on 500K dataset with varying minimum support

Algorithms (minimum support)	Memory usage (in MB) at varying minimum supports				
	20 %	10 %	9 %	8 %	7 %
MineHFPs	222	1150	2130	2770	3839
TidFP	78	530	722	crashed	crashed

Table 13. CPU execution time at minimum support of 10 % on varying sizes of dataset

Algorithms (dataset size)	Runtime (in Seconds) at different dataset sizes			
	125K	250K	500K	1M
MineHFPs	3311	8321	16233	34089
TidFP	10264	24008	48077	98858

of 20 %, 10 %, 9 %, 8 %, and 7 %). Table 13 describes the execution time of the MineHFPs and the TidFP algorithm at the minimum support of 10 % on dataset sizes of 125K, 250K, 500K, and 1M.

6   Conclusions and Future Work

More comprehensive and detailed real world data, such as different products on a Business to Customer (B2C) website, their histories, versions, price, images, or specifications are more suitable to be represented in an object-oriented database model. This paper proposes an object-oriented class model and database schema, and a series of class methods for mining multiple data sources. This paper also provides mechanisms that allow the flexibility of implementing this model with the popularly used relational DBMS. The methods can mine frequent patterns on each local object database and also mine the Hierarchical Frequent Pattern (MineHFPs) which specify at which hierarchy level the pattern is frequent in a global integrated table by extending Apriori-based TidFP algorithm. This paper also proposes object-oriented join (OOJoin) which joins superclass and subclass tables by matching their type and super type relationships. Thus, to implement the OO database model proposed using a relational DBMS, each relational database table corresponds to an OO class, each relation DB tuple corresponds to an OO class instance object. Each relational foreign key attribute is implemented with both the class type and supertype value of the class with the defined OOJoin condition. To improve the performance of the MineHFPs algorithm, this paper also extends map-gen join method used in TidFP algorithm to oomap-gen join for generating k-itemset candidate pattern to reduce the candidate generation and avoid unnecessary support counting by indexing the (k-1)-itemset candidate pattern using two position codes, start position and end position tied to inheritance hierarchy. The experimental results show that the proposed MineHFPs algorithm for mining hierarchical frequent patterns is approximately 3 to 4 times faster than the TidFP algorithm to mine the same patterns but have the trade off of costing 2 to 3 times more memory usage. However, the MineHFPs algorithm can discover the frequent pattern at different hierarchy levels in the format of $<$Tidlist, itemsets, $class_i>$. The TidFP algorithm can only discover the patterns in the format of $<$Tidlist, itemsets$>$. Our proposed object-oriented class model and database schema can be applied to other application domains, such as a Student Information System. Every department or faculty has its own database tables C_i. The Root table can be the class enrolment table and it may store the class and students enrolment information. The database tables C_i and Root do not include any historical attribute such as a time stamp (which may include date, month and year). Future work may include extending this model for representing and comparative analysis of non-structured multiple data sources such as documents, their derived forms (e.g., summaries), historical data sources (data warehouses), derived data (e.g., data warehouse materialized views). The historical attribute can display the history of the products and the history of sales transactions. Although the proposed object oriented data model representation of a database as presented in Sect. 2 currently focuses on a set of classes C_i connected by their class inheritance hierarchy H that is used to depict the superclass and subclass relationships between classes, this model is easily extendible to accommodate as well complex object type or attribute hierarchy where attributes are of type of another existing class. Current implementation and discussions assume all class attributes to be of simple type (e.g., string) and if attributes are of complex types (e.g., CPU is of type computer), they can be accommodated by having those complex attributes (e.g. CPU) as nested list of attributes (having all attributes of its complex type computer) and applying the process on all attributes including those inherited from the complex type (e.g., computer). Future work should extend the Mine-HFP algorithm to handle nested objects in the model definition such that as the model definition of inheritance hierarchy is provided, that of complex attribute hierarchy is also provided and an equivalent of OOJoin function for obtaining all inherited attributes of a complex attribute defined and used during both MineClassFP and MineRootFP methods.

References

1. Agrawal, R., Srikant, R.: Fast algorithms for mining association rules in large databases. In: 20th International Conference on Very Large Databases, Santiago, pp. 487–499. Morgan Kaufmann (1994)
2. Annoni, E., Ezeife, C.I.: Modeling web documents as objects for automatic web content extraction. In: ACM Sponsored 11th International Conference on Enterprise Information Systems (ICEIS), Milan, Italy, p. 91100. LNCS, Springer (2009)
3. Ayres, J., Flannick, J., Gehrke, J., Yiu, T.: Sequential patterns mining using a bitmap representation. In: ACM SIGKDD Conference, Edmonton, Canada, pp. 429–435. ACM (2002)
4. Ceci, M., Malerba, D.: Classifying web documents in a hierarchy of categories: a comprehensive study. J. Intell. Inf. Syst. **28**(1), 37–78 (2007)
5. Cheng, H., Zhou, Y., Yu, J.X.: Clustering large attributed graphs: a balance between structural and attribute similarities. ACM Trans. Knowl. Disc. Data **5**(2), 1–3 (2011)
6. Cheung, D., Ng, V., Fu, A., Fu, Y.: Efficient mining of association rules in distributed databases. IEEE Trans. Knowl. Data Eng. **8**(6), 911–922 (1996)
7. Dai, H.: An Object-oriented Approach to Schema Integration and Data Mining in Multiple Databases, pp. 294–303. IEEE Computer Society (1998)
8. Ezeife, C.I., Barker, K.: A comprehensive approach to horizontal class fragmentation in a distributed object based system. Int. J. Distrib. Parallel Databases (DPDS) **3**(3), 247–273 (1995)
9. Ezeife, C.I., Barker, K.: Distributed object based design: vertical fragmentation of classes. Int. J. Distrib. Parallel Databases (DPDS) **6**(4), 327–360 (1998)
10. Ezeife, C.I., Saeed, K., Zhang, D.: Mining very long sequences in large databases with PLWAPLong. In: 13th ACM Sponsored International Database Engineering and Applications Symposium, pp. 234–241. ACM (2009)
11. Ezeife, C.I., Mutsuddy, T.: Towards comparative mining of web document objects with NFA: WebOMiner system. J. Data Warehouse. Mining (IJDWM) **8**(4), 121 (2012)
12. Ezeife, C.I., Zhang, D.: TidFP: mining frequent patterns in different databases with transaction ID. In: Pedersen, T.B., Mohania, M.K., Tjoa, A.M. (eds.) DaWaK 2009. LNCS, vol. 5691, pp. 125–137. Springer, Heidelberg (2009)
13. Fortin, S., Liu, L.: An object-oriented approach to multi-level association rule mining. In: International Conference on Information and Knowledge Management, pp. 12–16. ACM (1996)
14. Han, J., Fu, Y.: Discovery of multiple-level association rules from large databases. In: 21st International Conference on very Large Databases, Zurich, Switzerland, pp. 420–431. Morgan Kaufmann (1995)

15. Han, J., Nishio, S., Kawano, H., Wang, W.: Generalization-based data mining in object-oriented databases using an object cube model. Int. J. Data Knowl. Eng. **25**(1), 55–97 (1998)
16. Han, J., Pei, J., Yin, Y., Mao, R.: Mining frequent patterns without candidate generation: a frequent-pattern tree approach. Int. J. Data Mining Knowl. Discov. **8**(1), 53–87 (2004)
17. Jin, Y., Murali, T.M., Ramakrishnan, N.: Compositional mining of multirelational biological datasets. ACM Trans. Knowl. Discov. Data **2**(1), 1–35 (2008)
18. Kemper, A., Moerkotte, G.: Object-oriented Database Management. Prentice-Hall Inc., Upper Saddle River (1994). ISBN: 0-13-629239-9
19. Kuba, P., Popelinsky, L.: Mining frequent patterns in object-oriented data. In: Proceedings of the 2nd International Workship on Mining Graphs, Trees and Sequences, ECML/PKDD, Pisa, pp. 15–25. University of Pisa (2004)
20. Satheesh, A., Patel, R.: Use of object-oriented concept in database for effective mining. Int. J. Comput. Sci. Eng. **1**(3), 206–216 (2009)
21. Sengupta, A.: On the feasibility of using conceptual modeling constructs for the design and analysis of XML Data. ACM Trans. Knowl. Discov. Data **72**, 219–238 (2012)
22. Wikepedia, The Free Encyclopedia. http://en.wikipedia.org/wiki/Object_database
23. Zaki, M.: SPADE: an efficient algorithm for mining frequent sequences. Mach. Learn. J., Special Issue on Unsupervised Learning **42**(1), 31–60 (2001)

Author Index